JAVA FOR BIOINFORMATICS AND BIOMEDICAL APPLICATIONS

JAVA FOR BIOINFORMATICS AND BIOMEDICAL APPLICATIONS

by

Harshawardhan Bal
Booz Allen Hamilton, Inc., Rockville, MD

and

Johnny Hujol
Vertex Pharmaceuticals, Inc., Cambridge, MA

Library of Congress Control Number: 2006930294

ISBN-10: 0-387-37235-0 e-ISBN-10: 0-387-37237-7
ISBN-13: 978-0-387-37237-8

Printed on acid-free paper.

Printed in the United States of America.

9 8 7 6 5 4 3 2 1

springer.com

Contents

Foreword

April 2006

Introduction

Bioinformatics is at a crossroads. We work in a field that is changing every day, increasingly moving from specific solutions created by single researchers working alone or in small groups to larger, often geographically dispersed programs enabled by collaborative computing and open software. This book represents an important development, giving the reader an opportunity to discover how the use of open and reusable Java code can solve large bioinformatics problems in a software engineered and robust way. I work with one of the authors of this book every day, on the National Cancer Institute's cancer Biomedical Informatics Grid (caBIG™) project, and I can attest that they are well suited to share with their readers both their experience in the development and use of bioinformatics software, as well as their interest in solid software engineering and interoperability.

Background and history

In its short history, bioinformatics has become an increasingly important part of how scientists involved in biological research go about their work. This has lead to an explosion of interest in the subject, and a similar explosion in tools and data resources for researchers to learn and use in their work. Historically, tools for bioinformatics have been idiosyncratic and are custom-developed by the end-users (or those close to them) in an iterative fashion until the specific immediate problem is solved. This has led to a balkanization of informatics systems, sometimes yielding multiple, incompatible systems at a single institution for a single application. This trend is beginning to change, with groups throughout the research community developing standards and shared data models, in areas ranging from gene expression arrays to pathways and proteomics. With a range of emerging software capabilities and a growing interest in interoperable tools and standards, bioinformatics practitioners have an ever-expanding toolbox from which to draw on to develop the basic software infrastructure behind their work. Similarly, with the increasing interest within the biomedical informatics community in the use of well-defined software engineering methodologies, and disciplines like design patterns and model-driven architecture, the software developed there will increasingly last longer, be easier to maintain, foster interoperability and reuse, and ultimately be more robust and cost effective.

Interfaces and standards

Interfaces and standards, as well as the use of well established development platforms, especially object-oriented programming, allow the bioinformatics practitioner to solve problems faster, with fewer lines of reusable, well-documented code than before. Through access to and study of well-established principles of software engineering and computer science, the solutions to problems in biomedical informatics will also be solid and optimally designed. With the increasing size of the datasets used in biomolecular informatics, derived from all manner of new high-throughput technologies and online databases, it is increasingly important to use thoughtful, efficient and well-established algorithms in the analysis of that data. Informatics students who can decompose complex, biologically significant informatics problems into simpler models, for which there are corresponding, validated and pre-existing software objects, will be amply rewarded for their efforts. It is by building on well-supported software platforms, using established and tested methodologies, that the most favorable balance can be achieved between effort and benefit.

Java as a platform

This book will teach you ways to make use of the Java programming language as a platform for your work in biomedical informatics, and in doing so, will open you up to the possibility of using a wide range of software objects in use throughout the large software engineering and computer science communities. Java is, of course, not the only object-oriented platform that is appropriate for bioinformatics. Perl is very well established, and are python, C++ and many others. The lessons that you can learn in Java are transferable to any object-oriented system, and Java is proving to be a solid platform for work throughout the informatics community. In the caBIG™ project that both Harshawardhan and I are a part of, Java is one of the main (but far from the only) programming languages used in that project. As a result, there is a lot of infrastructure available in the form of open-source code and open-content resources that are available for the busy researcher, serious student, or interested hobbyist. The latter chapters in this book detail how to connect with and make use of those resources to solve your own informatics programs.

The future

Through the efforts of a global community of biomedical informatics researchers, and through the prevalence of the Internet, it has become possible for any interested person to learn enough about biology, software engineering, and computer science, to contribute meaningfully to the emerging science of informatics. With the amount of openly available raw biological data growing by leaps and bounds every day, there is every reason to believe that you can contribute too, and the book that you hold in your hand is a great way to join in. Bon voyage!

Mark Adams
Program Manager
NCI Cancer Biomedical Informatics Grid (caBIG™)
Booz Allen Hamilton
Rockville, MD

Preface

On April 15, 2003, the International Human Genome Sequencing Consortium (IHGSC) – an association of laboratories from around the world which had jointly undertaken the Human Genome Project formally announced the completion of the colossal task they had set out to accomplish: the sequencing and assembly of the 3 billion bases that comprise the human genome. This was a truly landmark achievement for science and medicine. Today, the word "genome" has become a household term and together with bioinformatics has revolutionized how we approach biomedical research. The human genome project has led to identification of thousands of disease genes and paved the way for the development of newer drugs and treatments. Undoubtedly, the sequencing of the human and other genomes is just the beginning of the revolution that is unfolding right in front of our eyes. We are moving towards a paradigm shift in medicine, from just-in-time treatment that is given after the onset of symptoms to predictive and personalized treatment where the determination of the genetic factors predisposing an individual to disease is made right at birth and treatment started much before the onset of disease.

There is also a fundamental shift in how biomedical research is going to be conducted and funded in the years to come, especially, in areas such as cancer research and heart disease where there is a critical need to bring newer and better treatments for patients. Cancer has passed heart disease as the number one killer in UK and US and has been recognized by the World Health Organization as a major health problem across the globe. To meet this challenge, the US National Cancer Institute (NCI) has launched the biggest collaborative research program in 2003 called the cancer Biomedical Informatics Grid (caBIG™). In the words of NCI Director, Dr. Andrew von Eschenbach, "...caBIG will become the 'World Wide Web' of cancer research informatics and will accelerate the development of exciting discoveries in all areas of cancer research". Thus started the journey towards the NCI Challenge Goal, "To eliminate the suffering and death due to cancer by 2015" and together with it the efforts of more than 50 NCI-designated cancer centers, scores of research laboratories, Universities and public and private institutions across the country.

Where does J2EE come in the picture? The healthcare and medical research enterprise that we see today with its complex distributed Internet-enabled architecture is dependent on technologies that provide the critical infrastructure components necessary to fulfill its patient data safety, security and regulatory compliance requirements. Java has emerged as a powerful programming language for developing secure, scalable and robust web-enabled applications and is particularly well suited for building the many interrelated components of the geographically dispersed biomedical research and business engine. Together with support from a number of open source standards, J2EE offers a number of advantages for such applications and is the major platform for development efforts under caBIG™.

Why now?

We were confronted with this question early on in the writing of the book. The answer lies in the way the biomedical research enterprise has been transforming itself over the past decade or so and in doing so, promising to revolutionize the way we provide patient care. caBIG™ is based on the principles of open source, open access, open development and federation and uses J2EE and open source technologies for all software development efforts under the program. CaBIG™ is perhaps the next major landmark in the making in the history of biomedical research. Consequently, the time for a closer look at J2EE and open source technologies in a way that combines industry standard software engineering and design principles, genomics, bioinformatics and cancer research, is ripe.

This book is an attempt to fill that critical need. The main differentiating feature of the book is its focus on creating and integrating practical, useful tools for the scientific community in the context of real-life, real-value biomedical problems that researchers encounter on a routine basis. The book leverages technologies for molecular biology, genomics, bioinformatics, clinical research and cancer research developed by the National Cancer Institute Center for Bioinformatics (NCICB), the National Center for Biotechnology Information (NCBI, a division of the

National Library of Medicine (NLM) at the NIH), and scores of research organizations across the nation.

The book begins with an overview of the state of biomedical research today and the challenges it faces due to the silo model that has perpetuated over decades across universities and research centers across the world. It establishes a case for and the rationale behind the current move towards integrative, collaborative and standards based research platform through an introduction to the NCI caBIG™ program. It next provides an overview of emerging architectural trends such as Web Services and Service-Oriented Architecture. The book is not as much about the J2EE platform as it is about its *application* to building useful software and does not dwell on the theoretical aspects of the language or the platform; the authors (as well as the readers) recognize that several excellent works on that topic already exist. Instead the uniqueness of this book is that after just a short introduction, it takes a deep dive into demonstrating how to build highly functional graphical user interfaces for common and widely used bioinformatics tools that most researchers are familiar with and find indispensable for any kind of research activity. The reader is led through a step-wise and incremental software development approach with two goals in mind – to demonstrate a systematic standard software engineering approach to application development and, to activate a thoughtful design process in the mind of the developer that is aimed at exploring ways to enhance the functionality and usefulness for end-users. The applications that are considered the backbone of modern genomic and bioinformatics-driven research – Basic Local Alignment Search Tool (BLAST), Genscan gene prediction tool and others are used to illustrate this process. The reader will notice a significant amount of code in this book and realize that this is so by design. Although there are many ways of architecting a solution for a particular problem, we have illustrated one such approach while encouraging users to build their own. In doing so, we have also attempted to promote the reuse of tried and tested code from existing software libraries based on open source projects such as Apache, BioJava, caBIG™, and others.

Another differentiating feature of the book, best described by a reviewer, is we "...take a gradual and applied approach to combining Java and Bioinformatics". This statement, in fact, represents the very fabric of our strategy. By the same design, we have devoted little time on describing features and individual programming elements for which excellent and easily accessible documentation already exists. Our approach has also been to create pipelines where two applications are combined together along logical workflows that researchers normally use in their research environments to produce an enhanced application that has more utility than the individual applications.

The book does not profess to be the comprehensive tome on J2EE; instead, it is designed to cover a few of the important topics that lend themselves to use in the situations that are commonly encountered in this domain. It is hoped that a more focused approach would lead to a better and clearer understanding of the core capabilities of the platform than would be achieved by a lengthier treatment of the subject that cover all its different aspects. Indeed, the vastness and the complexity of the biomedical space and the pace and profundity with which science, technology, policy and legislation affect it is at times daunting. The authors acknowledge the challenge of writing on a topic this difficult and hope to address the concerns of the readers of this volume to identify gaps and produce a more inclusive title while providing time for the emerging technologies described in this book and others beyond the scope of this book to mature and gain wider acceptance by the user community.

With this background in mind, the book is especially tailored towards graduate students majoring in computer science, or information technology and who intend to take up careers in architecting software solutions for biomedicine and healthcare. It is also meant for practicing professionals who are actively involved in developing, maintaining or enhancing biomedical software and need to remain on the cutting edge of trends and standards in medicine and information. Finally, it will also be useful to molecular biologists, life scientists and clinicians who have a strong commitment towards understanding how software technologies can be put to use in solving the unique demands presented by the modern post-genomic translational research landscape.

This work would not be possible but for the many people who helped us get our thoughts together and organized to this point. We thank the many initial reviewers of this book who represent both private as well as public companies and research organizations including thought leaders in the field, many of whom are closely associated with the latest movements in information and biomedical technologies, and in their application to initiatives such as caBIG™. We thank Dr. Mark Adams, the caBIG™ Program Manager, for his wholehearted support for the book from concept to conclusion and for lending his expert insight into the future of biomedicine as captured in the Foreword for this book. We thank the good people at Springer – especially, Joseph Burns and Marcia Kidston and their team – for sticking with us throughout the process and coming to our assistance whenever we had the slightest of troubles. We also thank our individual families – the grown-ups (our wives) Nathalie Hujol and Snehal Bal, and not so grown-up (Arnav Bal, just 3 at the time of this writing), who knowingly or unknowingly – but by no means reluctantly – allowed us both to pursue this adventure and leave the life outside our small world for the better part of the 2005-2006 to flourish without our intercession for the most part.

To all our readers – whether you are an end-user or a developer, a biologist, a clinician or a bioinformatician or, indeed, one of the many documented cross-disciplinary "hybrid professionals" - we hope this book serves the small but meaningful purpose we began with in our minds and that it provides a vignette into the fast and exciting world of biomedical research. We value your feedback and will continue to incorporate your suggestions and work hard to meet your expectations in partnership with you throughout the lifetime of this book. We hope to hear from you!

Bon chance and bonne journee.

Harshawardhan Bal
Johnny Hujol

April 2006

Chapter I

Introduction to Bioinformatics and Java

The Origins of Bioinformatics

On April 15, 2003, the International Human Genome Sequencing Consortium (IHGSC) – the association of laboratories from around the world which had jointly undertaken the *Human Genome Project* (HGP) formally announced the completion of the project and the colossal task that lay at its core: the sequencing and assembly of the more than 3 billion bases that comprise the *Homo sapiens* (human) genome. This is a truly landmark achievement for science and medicine. According to Nobel Laureate James D. Watson, President of the Cold Spring Harbor Laboratory, "*The completion of the Human Genome Project is a truly momentous occasion for every human being around the globe.*" In the words of Elbert Branscom, Founding Director of the Joint Genome Institute (JGI), "*We will see everything before this like the dark ages of biology*".

The HGP has had wide ranging implications on every aspect of science and medicine. As a result of the HGP, scientists have mapped the DNA hieroglyphic of the human genome to an accuracy of 99.99 percent and have estimated that human life and all its molecular, cellular and organismal machinery is programmed by 30,000 odd individual genes. It has given birth to *Bioinformatics* - a new scientific discipline at the crossroads of biology, medicine and information technology and provided an impetus for the rapid development of the fields of *Genomics* (the study of the genome) and *Proteomics* (the study of the entire complement of proteins expressed by the genome). Along with the sequencing of the human genome, the sequencing of model plant and animal genomes such as *Arabidopsis thaliana* (thale cress), *Caenorhabditis elegans* (worm), *Danio rerio* (zebrafish) and *Drosophila melanogaster* (fruit fly) has led to the development of fundamentally new discovery approaches and technologies that promise to revolutionize medicine.

In the space of just a few years, we have taken a giant step closer to a paradigm shift from "just-in-time" medicine (where treatment is provided after the appearance of symptoms) to "predictive medicine" (where the entire spectrum of disease susceptibility of an individual can be mapped at birth and treated in advance of the appearance of disease). We are also moving closer to an entirely new concept in therapy - "personalized medicine" (as opposed to "generalized medicine"), where individuals receive treatment with "designer" drugs that are tailored to suit their specific genetic backgrounds, thereby maximizing therapeutic potential and minimizing the occurrence of adverse events.

Why does one person respond to a certain medication while another does not? Why do some women get breast cancer while others do not? Why are some individuals more susceptible to an infectious disease than others? These are the kind of questions that biologists are trying to address. The next few decades will be completely consumed in research that leads to answers to these issues. The need to analyze vast amounts of genetic data has lead to the growth of powerful technologies that enable researchers to study the regulation of tens of thousands of genes at the same time. To be able to perform these information intensive tasks, scientists and clinicians must be comfortable with both the biological and the computational aspects of Bioinformatics as well as with the basic tasks of retrieving, extracting, organizing, analyzing and representing the data. While *Perl* and other scripting languages are preferred for day-to-day analysis of biological data, they are not suited for creating enterprise-level software. A robust *Object-Oriented Analysis, Design and Programming* language such as *Java* is better suited for this purpose. The *Java 2 Enterprise Edition* (*J2EE*) framework provides the ability to develop distributed, multi-tier applications that can be deployed and connected over the web. J2EE is platform-agnostic, meaning that it can run on virtually any platform. This is because the Java code is compiled into an intermediate code called *byte code*, which is interpreted and executed by the *Java Runtime Environment* (*JRE*) at run-time. Since *JRE* is available on any platform, code once created in Java can be run on any operating system.

In this Chapter, we will explore some bioinformatics applications that have been written in Java in order to demonstrate the power of J2EE technologies for creating biomedical software. In particular, we will focus on applications that have been developed for cancer research that have achieved the "industry standard" reputation in modern research and are actively being integrated for use in such cutting-edge research initiatives as the National Cancer Institute's *cancer Biomedical Informatics Grid* program (caBIG™, http://cabig.nci.nih.gov/). In doing so, we will provide an introduction to caBIG™ in this chapter and discuss how the different tools and applications that are being built or are being brought under the caBIG™ umbrella are helping solve the many bottlenecks in biomedical research.

Current State of Biomedical Research

Traditionally, biomedical research has been (and is still being) conducted in laboratories around the world in relative isolation from other laboratories, even if the subject of research may have been (or is) the same. While this method of operation has over the decades led to a rich collection of research data and many significant biomedical discoveries, it has also led to the isolation of data and capabilities into independent silos of information and expertise that lie locked in databases or within people and inaccessible to the larger research community. In addition, since the majority of individual laboratories have evolved their own operating procedures, methodologies and vocabularies to suit their own specific research problems, there has been a relative dearth of standardized ways of conducting and reporting experimental data. The lack of standardization and data sharing has proven to be a significant impediment to biomedical research and directly affects our ability to design better and more effective treatments.

Experts all over the world now generally agree that a better use of research data, especially with the aim of enhancing the pace of biomedical research for the benefit of the patient, is through open collaboration and sharing. This approach eliminates duplication of effort and result in a more efficient use of limited resources. This realization is especially significant in the post-genomic era. Modern day *high-throughput* assay technologies have given researchers the power to probe living systems with unprecedented precision and depth. This has in turn led to the adoption of a "systems" approach to research with an increasing trend towards studying entire pathways, hundreds and thousands of genes and whole organisms in one single experiment. However, this approach has also led to an explosion of raw data. There is today an ever-increasing need to connect this raw data into meaningful actionable knowledge that can yield real insights into disease processes.

Another significant change is the realization that a more powerful way of conducting research is to integrate data from multiple different fields of study spanning basic (laboratory-based) and clinical (patient-focused) research. This new approach called "*Translational research*" requires a team approach between physicians, scientists, bioinformaticians, statisticians and a host of other professionals working closely together towards specific outcomes. This method of operation brings together the cellular, molecular, biochemical, genetic and other biological aspects of research together with a clinical understanding of disease that results in practical outcomes of valuable clinical relevance. For example, translational research on lung cancer may involve a team consisting of *molecular biologists*, *computational biologists* and *biochemists* on one hand and, *thoracic surgeons*, *medical oncologists*, *radiation oncologists* and *nurse practitioners* on the other to understand basic disease mechanisms and to improve patient outcomes.

The basic idea behind this approach is to assimilate as much corroborating evidence as possible to test and validate a hypothesis rather than dealing with separate isolated bits and pieces of raw data, which do not point to a robust testable hypothesis. With the appropriate standards, processes, policies and technologies in place, a researcher following a promising lead, for example, a gene or a protein that is significantly overexpressed in a specific cell population or in a laboratory model and is suspected to play an important role in disease causation, can extend the research in meaningful ways by:

1. performing experiments that prove that inhibiting protein overexpression or inhibiting a specific step in a biochemical pathway reverses the ill-effects of the abnormal protein expression or the aberrant pathway

2. confirming that the results can be duplicated in biospecimens - that is, samples derived from tissues obtained from specific human organs (for example, lungs) possessing the same disease pathology and characteristics, thereby extending the evidence in actual patient samples

3. confirming that the protein is not present in normal non-target tissues (for example, liver, kidney, etc.) to avoid occurrence of toxicity due to a chemical agent being tested for interventional therapy

4. identifying patient cohorts who fit the study criteria and conducting therapeutic clinical trials to test efficacy of known or experimental agents for interventional therapy

The over or under expression of a *biomolecule* (typically a gene or a protein) - that is, its presence in higher or lower amounts, respectively, under a diseased condition (as compared to the levels that are observed under normal conditions) is generally referred to as *differential expression.* The differentially expressed protein in question can serve as a signature or a fingerprint of the underlying disease mechanism and is the living system's response to an alteration in normal physiology caused by disease or other external stimuli. Since it is a signal or a "marker" with significant biological implications, it is called a *biomarker*. Biomarkers can be any biomolecule - proteins, peptides, nucleic acids, carbohydrates, lipids, metabolites, etc. - the concentrations of which may increase or decrease, under specific abnormal conditions. An example of a biomarker is cholesterol, which is commonly used to identify risk of heart disease. Biomarkers can be assayed by standard biochemical methods and can be used as indicators of disease states in diagnostics as well as provide targets for therapeutic intervention. The application of biomarkers to diagnostics includes the ability to diagnose and monitor disease, risk stratification, disease prognosis, drug eligibility, prediction of safety and efficacy, and therapeutic monitoring. The therapeutic aspect is equally important because they provide a reliable readout of drug function and treatment efficacy and therefore guide decisions on the clinical development of promising drug candidates.

The research can be further extended by identifying patient cohorts who fit the study criteria in clinical trials to test efficacy of known or experimental agents that inhibit overexpression or otherwise reverses the ill-effects of the causative protein. Of course, this is a rather simplistic representation of an actual research scenario. The researcher may spend months or even years studying disease causation in the laboratory eliminating other suspected causative agents, sifting through literature and accumulating data from studies performed by other scientists, mining the available data using statistical and analytic algorithms, and iterating through each of these steps till a model that fits the observed data can be created with a high-level of confidence. In reaching this goal, the researcher has to have access to the appropriate tools to identify relevant research, assure that the data can be compared across experiments done under different conditions or if not, apply the necessary manipulations using appropriate tools, have access to those tools, and have the necessary resources to identify tissues, experimental models or human subjects locally or at other institutions. Such "bench to bedside" research can be conducted only in a situation where data, resources, applications and people are connected with one another and accessible via standardized ways on a network or grid infrastructure. This is the rational and promise of NCI's caBIG™ program.

The cancer Biomedical Informatics Grid program

caBIG™ was started by the NCI in July 2003 as a pilot project to create a standards-based interoperable network of cancer centers across the nation to increase data sharing and cooperation between biomedical scientists and to enhance the pace of cancer research. The aim of caBIG™ is to integrate bioinformatics, *cancer informatics*, *tissue informatics*, and *pathology informatics* to create a network of data, applications and individuals who can share data and tools seamlessly across geographical boundaries. To cover the various aspects of the complex cancer research domain, caBIG™ is divided into four Workspaces - Clinical Trial Management Systems (CTMS), Integrative Cancer Research (ICR), In Vivo Imaging and the Tissue Banks and Pathology Tools (TBPT) Workspace. Simply stated, caBIG™ is putting the "e" in cancer research, leading to an "e-research" platform that integrates data and knowledge from basic (laboratory-based) research to clinical (patient-based) research. To draw an analogy with the term e-business that refers to the application of Internet technologies to streamline enterprise business processes, caBIG™ is aimed at building the infrastructure, processes and policies to make research data from multiple research centers available via the web, handle secure transactions across networks, support queries and secure information interchange between distributed institutions, and enhance the efficiency of the cancer research engine as a whole. Making cancer data available electronically over the Internet enhances the speed of access to information, offers the opportunity to globalize data access and interchange, enables access to the most up-to-date data, enables researchers to adapt and quickly incorporate the latest understanding of disease biology into their experimental designs, and ultimately, to respond faster to critical patient needs and provide high quality service.

While there are some parallels between biomedical research data and business data, the two differ fundamentally in many respects, especially with respect to data on patient related medical information. caBIG™ therefore has to create this e-research infrastructure in strict compliance with applicable federal regulations for the protection of what is known as *individually identifiable* health information that can be linked to personal medical data and, if exposed, provoke the risk of misuse. In particular, the privacy provisions of the *Health Insurance Portability and Accountability Act of 1996* (HIPAA), apply to and seek to protect patient health information that is created or maintained by health care providers who engage in certain electronic transactions, health plans, and health care clearinghouses. A detailed treatment of the HIPAA rule is beyond the scope of this book. Suffice to say that this federal law gives patients rights over any personal medical data that health professionals and care providers collect in medical records and sets rules and limitations around who can receive and view their personal health information.

caBIG™ Organization and Architecture

As of this writing, caBIG™ had grown to a large enterprise consisting of more than 70 individual projects, more than 800 individual participants spanning greater than 70 public and private organizations. The caBIG™ enterprise has to support a complex interplay of customers (patients, research investigators, clinicians, bioinformaticians, etc.), and federated data (both text and image), services and analytic tools (data extraction, organization, querying, mining, clustering and visualization tools) over the web, while ensuring that it meets the necessary performance and capacity requirements for such operations. By its very design, caBIG™ systems need to be compatible with other systems on the network and make data and services available irrespective of the type of web-based system or device accessing caBIG™ resources. The caBIG™ infrastructure has to provide fail-safe mechanisms to serve its resources in a continuous manner without downtime for optimal benefit for the research community. The need to access and distribute sensitive clinical, pharmacogenetic and financial billing data under caBIG™ means that appropriate technologies and policies must be implemented to assure privacy, confidentiality and integrity of data, while blocking unauthorized access. These are just a few issues that make the caBIG™ initiative such a complex undertaking. The NCI Center for Bioinformatics (NCICB) has a key role in the making of caBIG™ and is actively developing the critical infrastructure components needed to address these requirements. Information on a sampling of such tools, for example, the *Common Ontologic Representation Environment* (caCORE) Software Development Kit (caCORE SDK), the *Common Security Module* (CSM), *caAdapter* and others, can viewed at the NCICB website at the following URL (http://ncicb.nci.nih.gov/NCICB/infrastructure).

How does one design a secure and scalable solution for an enterprise this large that covers all the pieces – the biomedical and clinical organization, the computing infrastructure, including applications, systems, servers, storage and the network - of a complex and distributed modern research and healthcare environment? How can the various building blocks or business components be assembled to deliver the services and capabilities required to address the lifecycle needs of the federated biomedical enterprise? The presence of data, services and tools in a distributed manner and the requirement of data sharing between organizations via the web means that we can no longer develop monolithic applications with user interfaces that simply talk to a backend database. Instead, the architecture has to accommodate a new design consisting of several "layers" or "tiers" that may be present on separate physical machines, operate independently of one another and subserve specific functions. In effect, any number of such layers may be present and because of the functional separation that the layer architecture provides, each layer preserves its distinct identity and can be maintained without regard for the implementation details of other layers. In effect, this design affords the developer with immense convenience for use and maintainability because entire tier implementations can be modified without affecting the rest of the application. The users can in turn access the required resources in a seamless and transparent manner. Such an architecture is called an *n-tier architecture*. The n-tier architecture consists of several tiers that perform the following functions - the display or presentation of data, the conduct of business logic and the storage of data. These are commonly referred to as the *Presentation tier*, the *Business tier*, and the *Data* or *Persistence tier*, respectively. **Fig. 1.1** below provides a graphical representation of this model.

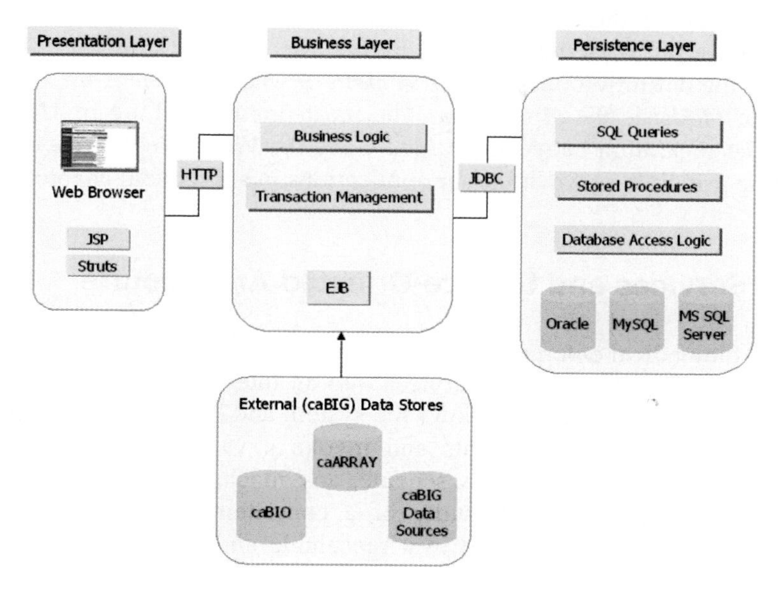

Fig. 1.1. Components of an n-tier architecture

The Model-View-Controller Framework

A concept that is closely associated with the n-tier architecture is a design principle called the *Model-View-Controller* (MVC) framework. The MVC framework defines separation between the data (Model), the visual component (View) and the communication that occurs between them (Controller). There are a number of advantages of using such a design.

The separation of components allows developers to prototype an application and validate its requirements quickly. The *view*, for example, can be designed and developed independently without affecting the design of the rest of the application. It's likely that the *View* will be modified more often than the *Model* (the data) to adapt to the requirements of users navigating through the user interface (UI). In addition, the way the *Model* is implemented is fully encapsulated and transparent to the other parts of the application.

The *Controller* handles the input that the *View* receives; it can then take action to update the *Model*. The *Controller* can also inform the *View* to update itself or the *View* can register itself as a *listener* of a *Model*, in which case the *View* will update anytime the *Model* notifies its *listeners*. This is the definition of the *observer pattern* where a *View* is the *Observer* and the *Model* is the *observable*. The most important thing in *MVC* is to keep the separation between the *Model* and the *View*. We will use this as a guiding principle as we build our applications in subsequent chapters.

Web Services and Service-Oriented Architecture

The biomedical enterprise needs to transform itself from an unorganized collection of data, tools and services into an interoperable, integrated and standards-based model that allows the system and its users to interact with a variety of business elements and invoke a variety of services along logical workflows. Under this scheme, any machine located on the web can be thought of as a provider of a consistent, reliable and defined "service" that can be invoked in a repeatable and standard manner. The *Basic Local Alignment Search Tool* (BLAST) server provided by the National Center for Biotechnology Information (NCBI), for example, provides a distinct service to a user – the ability to perform homology searches with a given nucleotide of amino acid sequence. The *Genscan* web server at MIT provides a different kind of service called "*gene prediction*" or the identification of complete gene structures in genomic DNA sequences. One can imagine the *World Wide Web* as made up of a large number of such services that can be accessed via standard Internet protocols such as *HTTP*, *FTP* etc. Each of these separate bits of functionality is a service and in each case, a service consumer (user or client) communicates and requests services from a service provider; the service provider in return communicates back the service requested. Both transactions (request and response) are carried out using messages that both parties can understand. Messaging between the services can be performed using the *eXtensible Markup Language (XML)*. This is the concept behind the emerging web architecture called *service-oriented architecture (SOA)*. The individuals services are connected together using *Web Services*, which define a set of technologies that enable connections between services.

The individual (web) services are self-contained, self-describing, modular applications that can be published, located and invoked across the

Web as well as discovered by other applications on the web. Each of these characteristics of a web service defines an essential component of the web services platform:

1. The means to communicate (pass messages and data) between services. This is usually achieved using *Simple Object Access Protocol* (*SOAP*), which defines a uniform way of passing XML-encoded data and a way to perform *remote procedure calls* (*RPC*s) using the Hypertext Transfer Protocol (HTTP) .

2. The ability to dynamically locate other services present on the web using a directory service. This is called *Universal Description, Discovery and Integration Service* (*UDDI*).

3. The ability to describe *what* a web service can do, *where* it resides, and *how* to invoke it. This is achieved through the *Web Services Definition Language* (*WSDL*).

As is apparent from the above, web services must use interfaces based on common Internet protocols such as HTTP and must use the XML standard for messaging. Although a detailed description of the web services platform and *SOA* is beyond the scope of this text, we will illustrate how the caBIG™ grid architecture called caGrid addresses the complex interoperability and integration issue we described earlier. We will delve into caBIG™ and the technologies being developed under the project in more detail in Chapter 6.

CaGrid

As mentioned briefly before, to make data interchange and collaboration possible, NCI and the caBIG™ participating institutions are using a number of technologies that the NCICB has been developing for the last several years. These include, for example, caCORE, Cancer Bioinformatics Infrastructure Objects (caBIO) and the Cancer Data Standards Repository (caDSR). These technologies allow integration of biomedical applications with a vast array of NCI data sources including genomic, animal model and clinical data. The NCI has also formulated compatibility guidelines to ensure that applications developed under the caBIG™ umbrella can interoperate with one another. The caBIG compatibility guidelines necessitate the use of controlled vocabularies and terminologies, Common Data Elements (CDEs), well documented API and *Unified Modeling Language* (*UML*) based object models to ensure interoperability with other caBIG applications. caCORE, which is caBIG's principle software development platform allows users to create caBIG™-compatible systems using an in-built modeling tool and a code generator.

The caBIG™ grid framework or *caGrid* is based on the service-oriented architecture model and open standards such as *Open Grid Services Architecture* (OGSA) created by the *Global GridForum* (GGF) for grid computing. The current version of caGrid as of this writing (caGrid 0.5) is built using the *Globus Toolkit 3.2* and the *OGSA Data Access Integration* (OGSA-DAI) framework version 5.0. The Globus Toolkit provides services and applications for the secure sharing and management of computing power, databases, and analytic tools over the web across organizational and geographic boundaries. OGSA-DAI component provides the middleware needed for accessing and integrating data via web services from the multitude of geographically distributed biomedical data sources on the grid including relational databases and XML based databases. Through the combination of these various components, caGrid empowers the caBIG™ engine and its users to develop and deploy of community provided services and API for building client applications.

Now that we have the basic background on caBIG™ and bioinformatics, lets examine a few software applications that are currently being used or are being developed under the caBIG™ program for oncology research to illustrate what scientists, clinicians, bioinformaticians and software engineers have together accomplished to address the needs in this area. We will use the research scenario we had discussed earlier - the differential expression of a gene and its product in a specific cell population or, in a disease model that leads to the plausible hypothesis that it has a role in disease causation - to provide examples of biomedical software applications. **Table 1.1** provides a breakdown of the translational research scenario into discrete sub-components and lists out the corresponding categories that apply to the scenario.

Table 1.1. Research use cases and corresponding categories

Research scenario	Category

Analyze genes that are differentially expressed in a specific cell population or a disease model	Gene expression analysis
Analyze proteins that are differentially expressed in a specific cell population or in a disease model	Proteomics
Analyze pathways that the differentially expressed molecules participate in	Pathway analysis
Query for and identify tissue samples located in distributed biospecimen resources that match the clinical, pathologic, and experimental parameters of the disease under investigation	Biospecimen inventory and annotation systems

Table 1.2 provides brief descriptions of the tools that we will introduce in this chapter to illustrate a representative set of Java-based bioinformatics applications. Also listed are the caBIG™ Workspaces under which each of the tools are being developed.

Table 1.2. Java-based bioinformatics tools

Name of application	CaBIG™ Workspace	Description
CaArray	ICR	Repository for managing, analyzing and visualizing of gene expression data from microarray experiments
CaWorkBench	ICR	Gene expression, pathway and sequence analysis, transcription factor binding site analysis, and pattern discovery
RProteomics	ICR	Statistical analysis, visualization m modeling of proteomics spectra
cPath	ICR	Integration and analysis system for integrating protein-protein interaction and molecular pathway information from multiple sources
caTissue Core	TBPT	Core biospecimen management tool for inventory, tracking and basic annotation of biospecimens.
CaTissue Clinical	TBPT	Tool for addition of pathology annotation to

Annotation Engine (CAE)		stored biospecimens using data from Anatomy Pathology systems, Clinical Pathology systems and tumor registries.
cancer Text Information Extraction System (CaTIES)	TBPT	Tool for extraction of pathology data such as tumor histology, staging, molecular markers, etc., from free text surgical pathology reports.

Let's look at each of the tools in turn and understand how they subserve or address a small component of the bigger research problem.

CaArray

caArray is an open-source standards-based repository for managing, analyzing and visualizing of gene expression data from microarray experiments. caArray enables researchers to make their microarray data publicly available to the larger cancer research community across geographically separated research centers via a web portal interface as well as through API. caArray uses a number of NCI technologies such as caCORE, caBIO and caDSR. In addition, caArray is built upon a number of caBIG™ compliant standards for data exchange such as Minimum Information About a Microarray Experiment (MIAME), MicroArray and Gene Expression Markup Language (MAGE-ML), MicroArray and Gene Expression Object Model (MAGE-OM) and uses controlled vocabularies based on the Microarray and Gene Expression Database (MGED) Ontology. caArray source code and API are available from NCICB for local installation under an open source license.

> MIAME is a set of guidelines that define the minimum set of data that is needed to enable the unequivocal interpretation of the results of a microarray experiment and to allow researchers the ability to reproduce the results of previously reported experiments. The guidelines include elements of microarray experiments such as aim and brief description of experiment, conditions under which the experiment was carried out, experimental design, quality control procedures used, the experimental protocol used, protocol and conditions used for hybridization and processing of the array, data normalization, extraction and processing protocols, etc.

The MicroArray and Gene Expression (MAGE) group aims to provide a standard for the representation of microarray expression data that would facilitate the exchange of microarray information between different data systems. This is being done under the aegis of the Object Management Group™ (OMG™), an international not-for-profit consortium defining standards for distributed object computing and interoperable enterprise applications. This has led to the establishment of a data exchange object model (MAGE-OM) and data exchange format (MAGE-ML) for microarray expression experiments. The purpose of the MGED Ontology is to provide standard terminology for the annotation of microarray experiments and to enable unambiguous descriptions of how the experiment was performed.

caArray is available for download at the NCI website at the following URL: http://caarray.nci.nih.gov/. **Fig. 1.2** shows the outcome of a query run on the caArray web portal for an experiment performed by investigators on the classification of complex diseases such as Diffuse large B-cell lymphoma to identify targets for interventional therapy.

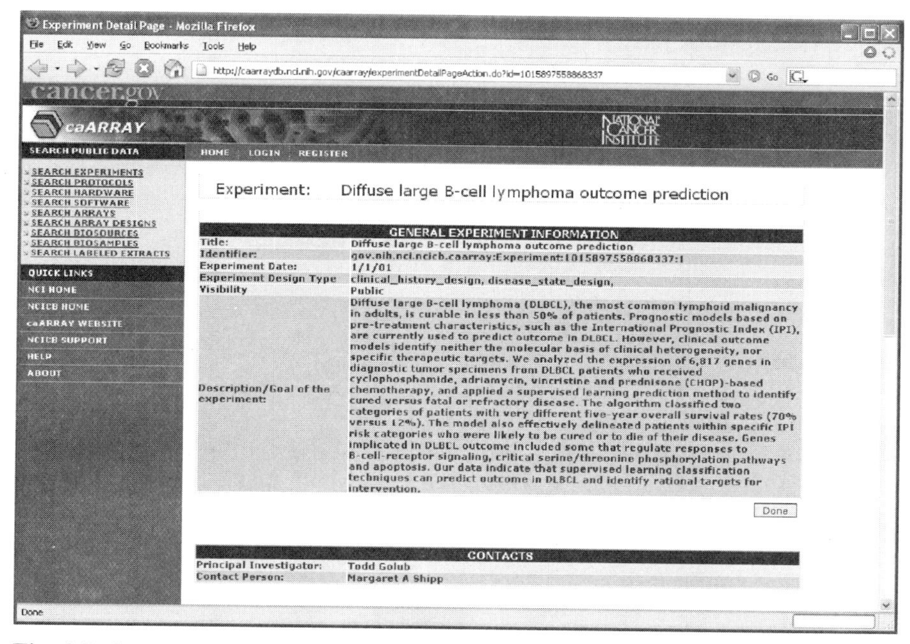

Fig. 1.2. Querying the caArray web portal for information on Experiments

Fig. 1.3 shows the results of a query to identify frozen samples (Biosource type) of type "cell" with name "lung" for organism "Homo sapiens" (that is, human samples) supplied by NCI.

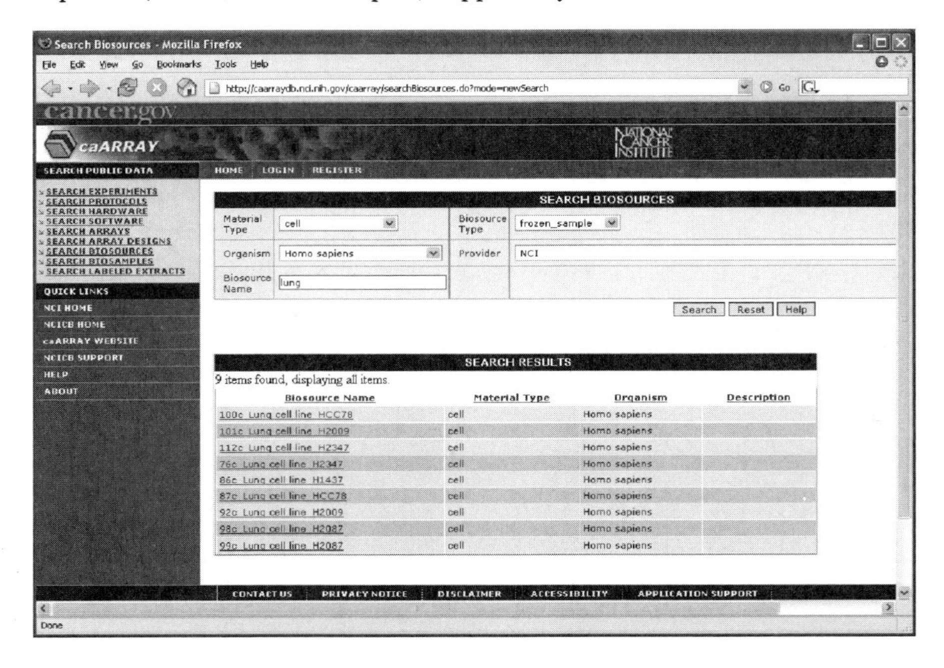

Fig. 1.3. Querying the caArray web portal for information on Biospecimens

CaWorkbench

caWorkbench is a suite of tools for loading, visualizing and analyzing gene expression data and provides the capability to integrate data of different types and from across a number of research institutions. caWorkbench is written with the Java programming language, uses the Java SWING libraries for creating the user interface. It runs on any platform that supports Java 1.5 including Windows XP, Solaris, Linux and OS X 10.5. The software is built on a component based architecture where each feature within the application such as pathways, annotation, expression profiles, etc. is available as a separate component that can be loaded individually when the application is started. caWorkbench is designed to retrieve data from the caArray database via the MAGE-OM API, and utilizes NCICB's caBIO API to access genomic, cancer models, molecular pathway and clinical trials information. caWorkbench is available for download from the NCI website at http://ncicb.nci.nih.gov/download.

RProteomics

The goal of the RProteomics project is to build open-source tools and develop standards for proteomics data analysis. As described earlier, Proteomics is the systematic study of the complete complements of proteins expressed by the genome. While gene expression is a study of the process of gene transcription (the synthesis of RNA from DNA), proteomics is the study of the process of gene translation (the synthesis or expression of protein from RNA). The protein machinery constitutes the signal transduction mechanism of a living cell or organism and is responsible for much of the physiological processes that sustain life. Proteomics is therefore a powerful tool in the arsenal of the biologist in the pursuit of molecular mechanisms of disease. Proteomics encompasses the determination of protein expression levels, protein-protein interactions, protein localization, and regulation by post-translational modifications, etc., ultimately to decipher protein function. The basic methodology in proteomics is the separation of proteins in a sample by gel electrophoresis, extracting the proteins of interest and followed by mass spectrometry (MS) to determine their identity and characteristics.

RProteomics derives its name from the open source R software environment that it uses for statistical analyses and visualization of proteomics data. In the future it will also provide a proteomics repository and access to proteomics data via web services. RProteomics includes statistical routines to analyze spectrometric data including algorithms for background curve determination, denoising, peak calibration, normalization of peak intensities, and predictive modeling. RProteomics supports the *mzXML* proteomics data standard and the *MIAPE* (*Minimal Information About a Proteomics Experiment*) standard, the latter of which is being developed by *The Human Proteome Organisation Proteomics Standards Initiative* to standardize data representation in proteomics and facilitate data comparison and exchange.

cPath

The cBio Pathway Information Resource or cPath is an open source pathway integration and analysis system for integrating protein-protein interaction and molecular pathway information from multiple sources. It also provides data visualization and analysis functionality via Cytoscape, another open source platform for visualizing interaction networks and integrating them with gene expression profiles. CPath provides access to data via a standard web service query interface that connects with a MySQL database backend as well as a HTTP based web service. Java and is based on a *3-tier architecture* using Java *servlets* and *Java Server Pages* (*JSP*). We will learn more about the Java servlets and JSP technology in chapter 4. Briefly, servlets and JSP provide a server and platform independent mechanism to create web-based applications that can serve dynamic web content.

CaTissue Core, caTissue Clinical Annotation Engine and caTIES

The simple research scenario we outlined earlier assumes that researchers can locate the biospecimens or tissues samples with the matching disease pathology or disease parameters so to perform the necessary follow-up and validation studies. For example, researchers may want to query a database for biospecimens that have associated gene expression data for a gene or set of genes that may be differentially expressed under a specific disease condition. Under caBIG™, the functionality to manage, annotate and identify matching biospecimens that may be present in a federated manner in geographically dispersed research institutions is being done through the caTissue suite of tools - caTISSUE Core (not to be confused with caCORE), caTissue Clinical Annotation Engine and caTIES.

These are some of the most advanced tools that are currently available in caBIG™ in terms of the software development effort as well as in terms of their adoption by a number of cancer centers and research institutions across the nation. We will illustrate the development efforts behind the caTissue Core application to demonstrate how the various elements of the J2EE platform have been applied to create a robust application to facilitate tissue banking operations.

CaTissue Core

As described earlier, the function of the caTISSUE Core system is to serve as the base or core solution for biospecimen inventory, tracking and basic annotation for use across cancer centers and other institutions with biospecimen resource facilities. In addition, CaTissue Core establishes the foundation of the TBPT object model that represents the tissue banking and pathology domain. Together with the other TBPT applications – caTissue Clinical Annotation Engine and caTIES, caTissue Core constitutes what is called the caTISSUE system, the comprehensive suite of tools for managing the life cycle events and operations of the tissue banking and pathology information domain.

The caTISSUE Core application is comprised of an n-tiered architecture (**Fig. 1.4**). The presentation tier consists of a web interface as well as HTTP based Java API. The web application used Java Server Pages (JSP) technology to serve dynamic web content. The HTTP API enable users to access all caTissue Core functionality that is available through the web based application. The web-based user interface is designed using the Apache Struts framework following the Model-View-Controller (MVC) Model 2 design approach. The Model 2 approach is a variation of the classic Model-View-Controller (MVC) design paradigm we described earlier. Applied to the Java servlet and JSP technology, under Model 2, the execution of the business logic is managed by the servlet and the presentation logic is managed by the JSPs. CaTissue Core also uses the Tiles framework which specifies the layout of each JSP page using templates and provides a mechanism to manage and reuse the various visual components such as the headers, footers and navigational elements of individual web pages. The caTissue Core business tier contains domain objects and model classes where the tissue banking related business logic resides. The Persistence tier is a local database for storage of tissue banking data, as well as, external data sources such as NCI's Cancer Data Standards Repository (caDSR) and Enterprise Vocabulary Services (EVS).

CaTissue Core provides two mechanisms for interaction between the user interface and the backend data stores - through an Object-Relational Mapping (ORM) tool called *Hibernate* and through the Java Database Connectivity (JDBC) API. Hibernate is used to define the mapping between Java classes to the tables in a relational database in order to persist the objects in a relational database. JDBC API provide database-independent connectivity and access to a wide range of SQL databases as well as other types of data sources, including spreadsheets and flat files. caTissue Core provides support for Oracle as well as MySQL databases.

The caDSR and EVS are a set of resources and tools to describe biomedical data and concepts in standardized ways using Common Data Elements (CDE) and controlled vocabulary, respectively. Access to these services is provided through the caCORE API. We will learn more about these resources in Chapter 6.

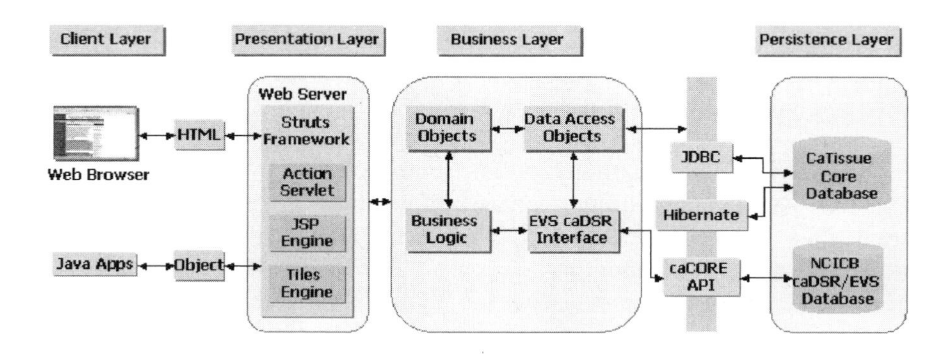

Fig. 1.4. caTissue Core n-tier architecture

Summary

This Chapter provides a brief introduction to The Human genome Project, perhaps the single most important event in the history of medicine after the elucidation of the double-helix structure of the DNA and to the fields of Bioinformatics, Genomics and Proteomics. While computing technology and software have played a fundamental role in the advancements that medical research has made in the last few decades, they have also led to problems in data quality. The silo approach that the biomedical research enterprise has taken has led to isolation of critical scientific expertise and knowledge, depriving patients of the benefits of modern science. To correct these issues, and to bring speedier benefits to individuals with cancer, the NCI in partnership with its Center for Bioinformatics and a number of Cancer Centers across the country launched the caBIG™ program with the aim of providing scientists with the infrastructure and resources to better control, share, assimilate and integrate data from disparate sources. The chapter also provides an overview of the role that the J2EE platform has played in biomedical research especially with the advent of the Internet age and the availability of the WWW as a catalytic medium for the sharing of resources across space. We also provided examples of a few software applications that demonstrate the power of the J2EE platform

In the next Chapter, we will build on the understanding we have gained so far of the state of and the challenges faced by the biomedical enterprise and begin the exercise of understanding how software is built using the J2EE platform. We will illustrate this by building an application using the Java *Swing* library to run biological sequence searches using the NCBI BLAST engine.

Questions and Exercises

1. Trace the origins of the Human Genome Project beginning from the elucidation of the structure of DNA in 1953. What were some of the landmark events and technologies associated with the successful sequencing of the human genome?

2. Visit the caBIG™ website to learn more about its organization and activities. Identify the main reasons behind the launch of the caBIG™ project. What are the technological and social hurdles that caBIG™ has to overcome in order to be successful? How will caBIG™ transform medicine if it meets its goals?

3. Compare HGP and caBIG™. What are some of the parallels you can draw between the two projects? Think about how these projects contribute to understanding of disease, especially cancer, and the advancement of modern medicine.

4. What tools and technologies are being created by the NCICB and participating cancer centers to advance the caBIG™ mission? What role does J2EE play in this effort?

Additional Resources

- Apache Struts - http://struts.apache.org/index.html

- caBIG™ Compatibility Guidelines - http://cabig.nci.nih.gov/guidelines_documentation
- caDSR - http://ncicb.nci.nih.gov/NCICB/infrastructure/cacore_overview/cadsr
- CaWorkBench - http://wiki.c2b2.columbia.edu/workbench/index.php/Main_Page
- EVS - http://ncicb.nci.nih.gov/NCICB/infrastructure/cacore_overview/vocabulary
- Genscan - http://genes.mit.edu/GENSCAN.html
- Global GridForum - http://www.gridforum.org/
- Hibernate - http://www.hibernate.org/
- HGP (US Department of Energy site) - http://doegenomes.org/
- MAGE-ML - http://www.mged.org/Workgroups/MAGE/mage-ml.HTML
- MGED Ontology - http://mged.sourceforge.net/ontologies/MGEDontology.php
- NCBI BLAST - http://ncbi.nih.gov/BLAST/
- OGSA - http://www.globus.org/ogsa/
- OMG - http://www.omg.com/
- The OGSA-DAI project - http://www.ogsadai.org.uk/
- Unified Modeling Language - http://www.uml.org

Selected Reading

Initial sequencing and analysis of the human genome. Lander et al. Nature. 2001 Feb 15;409(6822):860-921.

The sequence of the human genome. Venter, JC et al. Science. 2001 Feb 16;291(5507):1304-51.

The caCORE Software Development Kit: streamlining construction of interoperable biomedical information services. Phillips J, Chilukuri R, Fragoso G, Warzel D, Covitz PA. BMC Med Inform Decis Mak. 2006 Jan 6;6:2.

Covitz PA, Hartel F, Schaefer C, De Coronado S, Fragoso G, Sahni H, Gustafson S, Buetow KH. caCORE: a common infrastructure for cancer informatics. Bioinformatics. 2003;19:2404–2412.

Common data element (CDE) management and deployment in clinical trials. Warzel DB, Andonaydis C, McCurry B, Chilukuri R, Ishmukhamedov S, Covitz P. AMIA Annu Symp Proc. 2003; 1048.

Chapter II

Introduction to Basic Local Alignment Search Tool

The Basic Local Alignment Search Tool or BLAST, as it is commonly referred to as, is a database search tool, developed and maintained by the National Center for Biotechnology Information (NCBI). The web-based tool for BLAST searches is available at:

`http://www.ncbi.nlm.nih.gov/BLAST/`

The BLAST suite of programs has been designed to find high scoring local alignments between sequences, without compromising the speed of such searches. BLAST uses a heuristic algorithm which seeks local as opposed to global alignments and is therefore able to detect relationships among sequences which share only isolated regions of similarity (Altschul et al., 1990). The first version of BLAST was released in 1990 and allowed users to perform ungapped searches only. The second version of BLAST, released is 1997, allowed gapped searches (Altschul et al., 1997).

The Purpose of BLAST

Why is BLAST so useful for biologists? It is not uncommon nowadays, especially with the large number of genomes being sequenced, that a researcher comes across a novel DNA or protein sequence for which no functional data is available. Some basic information on the sequence is necessary before a molecular biologist can take the new sequence into the laboratory and perform meaningful experiments with it. It would, for example, make the task of deciphering the biological function of a piece of DNA much easier if it were known that the new sequence encoded a *metabolic enzyme* or, indeed, a protein that is a putative member of a *superfamily* such as an *immunoglobulin*, a *kinase*, etc. Conversely, if the sequence was a Repetitive DNA Element, it would need an entirely different approach for its study.

This is where the power of database searching comes in handy. The principle aim of database searching, in general and with BLAST, in particular, is to reveal the existence of similarity between an input sequence (called 'query sequence') that a user wants to find more information about and other sequences (called 'target' sequences) that are stored in a biological database. This is usually the first step a researcher takes in determining the biological significance of an unknown sequence.

Given the size of biological sequence databases maintained by NCBI (the non-redundant set of sequences were estimated at 540 million residues in 2004), database searches usually reveal sequences that have some degree of similarity to the query sequence. These sequences from the database that come up with similarities with the input sequence are commonly referred to as 'hits'. Once such hits are found, users can draw inferences about the putative molecular function of the query sequence. A thumb rule for drawing inferences is that two sequences that share more than 50 per cent sequence identity are usually similar in structure and function. Under such conditions, the major sequence features of the two sequences can be easily aligned and identified. If there is only a 25 per cent sequence identity, there may be some structural homology, although in such situations, the domain correspondence between the two proteins may not be easily apparent. It is also generally accepted that sequences that are important for function (and therefore, for the survival of an organism or species) are generally conserved.

An example where a database search resulted in an important discovery was the finding reported by Doolittle et al. (1983) of the similarity between the *oncogene*, *v-sis*, of *Simian sarcoma virus* (an *RNA tumor virus*) and the gene encoding *human platelet-derived growth factor* (PDGF). The v-sis gene was the first oncogene to be identified with homology to a known cellular gene. This discovery provided an early insight into the critical role that growth factor signaling plays in the process of malignant transformation. Another example of the value of database searching was the discovery that the defective gene that caused *cystic fibrosis* formed a protein that had similarity to a family of proteins involved in the transport of hydrophilic molecules across the cytoplasmic membrane (Riordan, et. al., 1989). Cystic fibrosis is the most common inherited disease in the Caucasian population and affects the respiratory, digestive and reproductive systems. It is now known that mutations in the cystic fibrosis gene lead to loss of chloride transport across the cell membrane, which is the underlying cause of the disease.

Performing a BLAST Analysis

Before we can build a BLAST application, we need to understand how BLAST searches are performed using the NCBI BLAST service. BLAST is actually a suite of programs – the particular choice of program(s) depends on the type of input sequence (amino acid or nucleotide) and the type of the database to be searched against (protein or nucleotide). The most commonly used search programs and their applications are described in **Table 2.1.**

Table 2.1. BLAST programs

Program	Comparison	Application
BLASTN	DNA vs. DNA. Compares a nucleotide query sequence against a nucleotide sequence database.	Find DNA sequences that match the query
BLASTP	Protein vs. Protein. Compares an amino acid query sequence against a protein sequence database.	Find identical (homologous) proteins
BLASTX	DNA vs. Protein. Compares a nucleotide query sequence translated in all reading frames against a protein sequence database.	Find which protein the query sequence codes for
TBLASTN	Protein vs. DNA Compares a protein query sequence against a nucleotide sequence database dynamically translated in all reading frames.	Find genes in unknown DNA sequences
TBLASTX	DNA vs. DNA Compares the six-frame translations of a nucleotide query sequence against the six-frame translations of a nucleotide sequence database.	Discover gene structure. (Find degree of homology between the coding region of the query sequence and known genes in the database.)

In summary, the available BLAST options are:

1. For nucleotide sequences: BLASTN, BLASTX and TBLASTX

2. For amino acid sequences: BLASTP and TBLASTN

In the simplest case, we need the following pieces of information to perform a BLAST search using NCBI's web-based service (http://www.ncbi.nlm.nih.gov/BLAST/):

1. An input query sequence (this can be a nucleotide or amino acid)

2. The database to search against (this can be a nucleotide or protein database)

3. A database search program (any of the five available BLAST options)

Additional parameters such as the matrix and E-values also need to be set. Once the user submits the necessary information, the BLAST engine responds with a message informing the user that the request has been successfully submitted and placed in a queue. The server also provides an estimate of the time in which the results will become available for viewing. The BLAST output itself consists of a header that provides information on the specified BLAST parameters, the request ID for the search, the length of the query sequence and the database used. **Fig. 2.1 - 2.3** show the results immediately after initial submission of and the output of a BLAST search performed with the *human cystic fibrosis transmembrane conductance regulator* (CFTR) mRNA sequence (gi: 90421312). **Fig. 2.1** and **Fig. 2.2** show the request ID (RID) that uniquely identifies this particular search job that was submitted to the BLAST queue. We will learn more about RID in Chapter 3 when we build the functionality to perform BLAST searches using the NCBI QBlast service. **Fig. 2.2** provides a view of the header information present in the BLAST search results.

Below the header is a line up of sequences from the selected database ("hits") that match the query sequence along with the number of matches found (**Fig. 2.3**). A mouse-over on the first line reveals information on the origin of the sequence (for example, whether it is a human or a mouse sequence, the name of the gene, if known) and the score (**Fig. 2.4**). Sequences on the top are more significant (have better matches to sequences in the database and thus, have higher scores) than those at the bottom (lower scores).

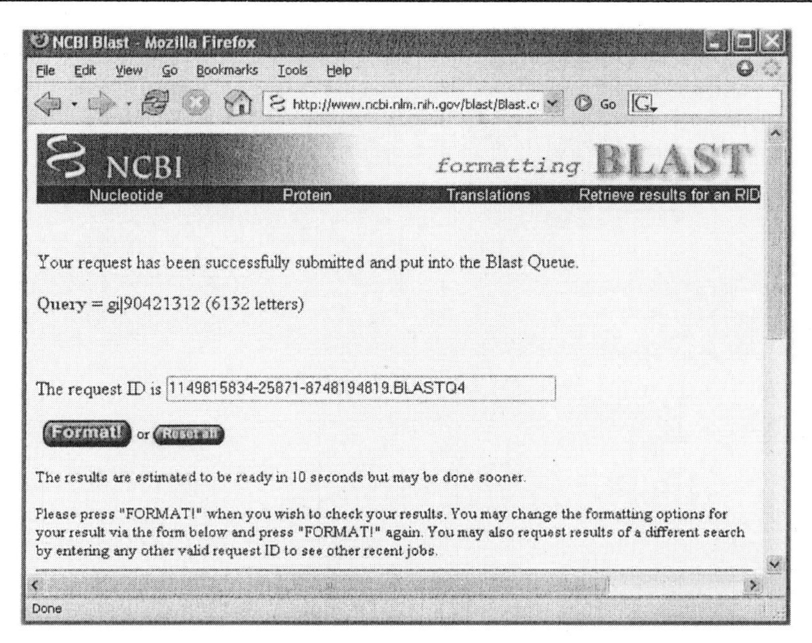

Fig. 2.1. Submission of a sequence to the BLAST queue

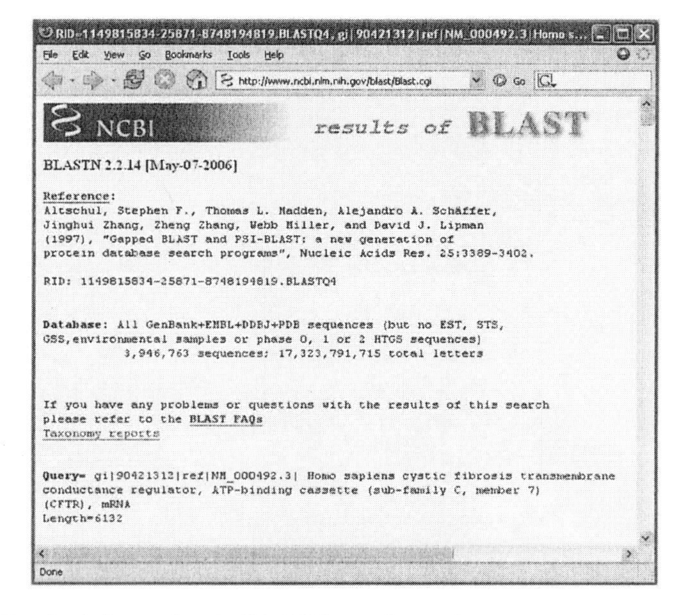

Fig. 2.2. Header information in BLAST search results

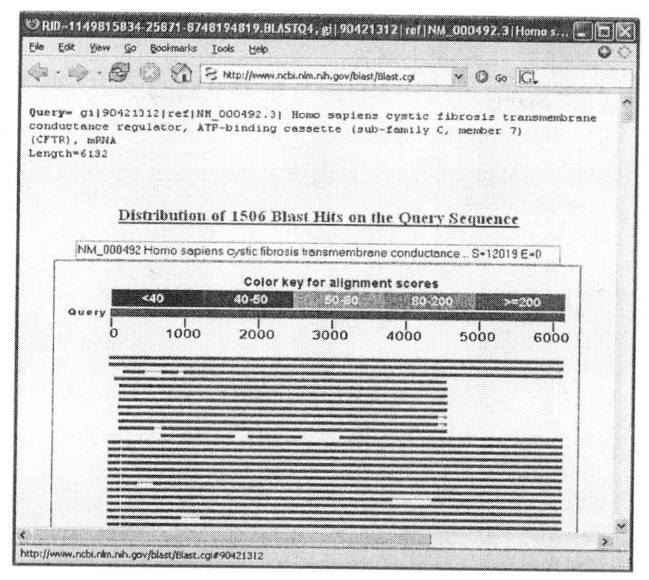

Fig. 2.3. Alignment of BLAST hits to the input sequence

RID=1149815834-25871-8748194819.BLASTQ4, gi|90421312|ref|NM_000492.3|Homo sapiens cystic fibro...

File Edit View Go Bookmarks Tools Help

http://www.ncbi.nlm.nih.gov/blast/Blast.cgi

Sequences producing significant alignments:	Score (Bits)	E Value	
gi|90421312|ref|NM 000492.3| Homo sapiens cystic fibrosis tra...	1.202e+04	0.0	
gi|180331|gb|M28668.1|HUMCFTRM Human cystic fibrosis mRNA, en...	1.194e+04	0.0	U E G
gi|55629259|ref|XM 519330.1| PREDICTED: Pan troglodytes simil...	9791	0.0	G
gi|47933787|gb|AY608405.1| Cysteine-free hCFTR in pGEMHE, comple	8199	0.0	
gi|74136424|ref|NM 001032938.1| Macaca mulatta cystic fibrosi...	8179	0.0	G
gi|3047170|gb|AF013753.1|AF013753 Macaca mulatta cystic fibro...	8179	0.0	G
gi|46452254|gb|AY585334.1| Sus scrofa cystic fibrosis transme...	5461	0.0	U
gi|6007842|gb|AF189720.1|AF189720 Oryctolagus cuniculus chlor...	5241	0.0	
gi|55742781|ref|NM 001007143.1| Canis familiaris cystic fibro...	5027	0.0	U G
gi|54873161|gb|AY780429.1| Canis familiaris cystic fibrosis t...	5027	0.0	U G
gi|1100984|gb|U40227.1|OCU40227 Oryctolagus cuniculus CFTR chlor	4601	0.0	
gi|1669376|gb|AC000061.1| Homo sapiens BAC clone CTB-133K23 from	3433	0.0	E
gi|89348170|gb|DQ388142.1| Homo sapiens isolate cftr11589_A c...	3433	0.0	
gi|89348179|gb|DQ388145.1| Homo sapiens isolate cftr13838_B c...	3433	0.0	
gi|89348164|gb|DQ388140.1| Homo sapiens isolate cftr11521_A c...	3433	0.0	
gi|89348161|gb|DQ388139.1| Homo sapiens isolate cftr11376_B c...	3433	0.0	
gi|89348146|gb|DQ388134.1| Homo sapiens isolate cftr11321_A c...	3433	0.0	
gi|89348140|gb|DQ388132.1| Homo sapiens isolate cftr10976_A c...	3433	0.0	
gi|86169633|gb|DQ356260.1| Homo sapiens isolate cftr10975_B c...	3433	0.0	
gi|85724377|gb|DQ354391.1| Homo sapiens isolate cftr01814_B c...	3433	0.0	
gi|89348176|gb|DQ388144.1| Homo sapiens isolate cftr13838_A c...	3429	0.0	
gi|89348173|gb|DQ388143.1| Homo sapiens isolate cftr11589_B c...	3426	0.0	
gi|89348158|gb|DQ388138.1| Homo sapiens isolate cftr11376_A c...	3424	0.0	
gi|89348155|gb|DQ388137.1| Homo sapiens isolate cftr11373_B c...	3424	0.0	
gi|89348149|gb|DQ388135.1| Homo sapiens isolate cftr11321_B c...	3424	0.0	
gi|89348143|gb|DQ388133.1| Homo sapiens isolate cftr10976_B c...	3424	0.0	

Done

Fig. 2.4. Definition of database hits

Developing the SwingBlast Application

Now that we understand the significance and the working of the BLAST engine, we can begin our journey into the world of Java development by building a BLAST application, which we will call *SwingBlast*, from the ground up. In this Chapter, we will create the user interface elements using *Java Foundation Classes* or *JFC*, also known as *Abstract Windowing Toolkit (AWT)* and *Swing* classes. In Chapter 3, we will write the actual code to run the BLAST searches based on the NCBI BLAST engine. In each case, we will build the application in an iterative fashion thereby demonstrating a step-wise approach to building software - creating a basic program structure or framework and adding bits of code in an incremental fashion to enhance its functionality.

The steps for building Java applications from a software engineering point-of-view are as follows:

1. Develop use case scenarios
2. Define software modules
3. Define classes
4. Write the Java code (business logic)
5. Run and analyze output

We will begin by creating *use cases* that define the actions that a user may wish to perform on the application and the behavior that a user expects from the application in response to those actions. Use cases, simply stated, are individual scenarios that allow software developers to layout the behavior and functionality expected of the software. To create a Java based BLAST application that allows users to submit sequences and to retrieve the results of the search operation, we can envision the following use case scenarios:

1. User provides input information to the application
2. User submits the input information to the NCBI BLAST server

3. The application displays the selected BLAST results in graphical format

Fig. 2.5 provides a UML diagram that describes the interactions between the user and the application. The specific details about the expected input and output are as follows:

1. User provides input information to the NCBI BLAST engine: The input data can be a sequence or, if available, the corresponding sequence id from GenBank® (an annotated repository of all publicly available DNA sequences maintained by the NIH), which uniquely identifies a sequence within the GenBank database. The application behavior in either case is as follows:

   a. The input information is a nucleotide or protein sequence: In this case, after the sequence information is provided, the application automatically recognizes the sequence type, loads it in the Fasta format (**Fig. 2.6**) and presents the appropriate valid BLAST options (for example, BLASTN for nucleotide and BLASTP for protein etc., as explained in **Table 2.1**). The invalid BLAST options are disabled.

   b. The input information is a valid GenBank id (also called the GI number). In this case, the application downloads the sequence from GenBank and displays it in the appropriate format as stated above.

2. User submits the sequence to the NCBI BLAST server. Once the sequence becomes available to the application (either directly supplied by the user or downloaded from the GenBank id), the user selects the necessary BLAST parameters (the type of BLAST program, the database, the matrix, the E values, etc.) and hits the "Submit" button. This sends the sequence to the NCBI BLAST server for the search operation.

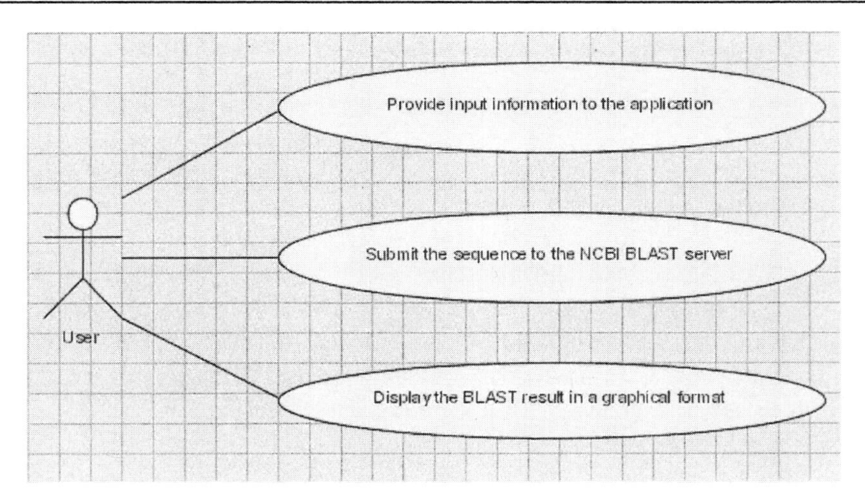

Fig. 2.5. UML diagram for the SwingBlast use cases

The last use case ("User wants to browse the BLAST results in a graphical format") arises from a need to view the BLAST output, that is, the list of sequences from the database that matched the input sequence in a graphical and interactive fashion.

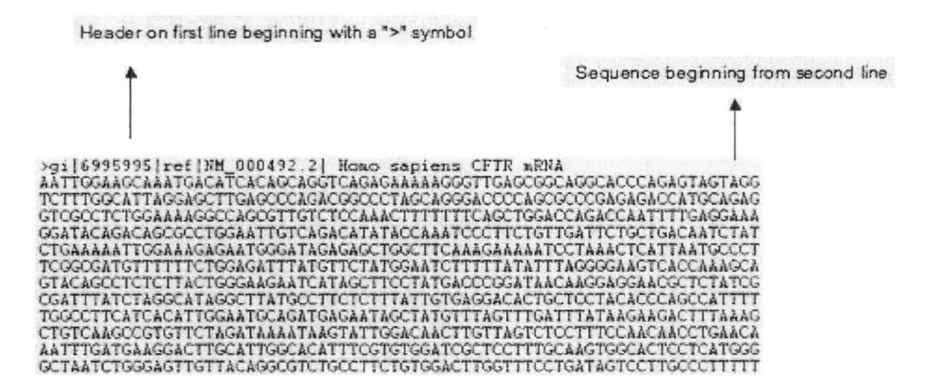

Fig. 2.6. A sequence represented in Fasta format

Designing the SwingBlast Java Application

The `SwingBlast` application involves data input from the user (the sequence or the GI number which identifies the sequence), manipulation of the input data ("BLASTing" the sequence against the selected databases), and visualization of the results of the database search (the BLAST output). Clearly, there are different parts to the application each of which performs a different function. We will follow the MVC framework we described in Chapter 1, while designing the various pieces of functionality of the `SwingBlast` application.

In line with the incremental approach to building the `SwingBlast` application, we will as a first step, create the basic framework application that will perform two basic functions - allow users to input a nucleotide sequence to the application and to format it in the Fasta format. The structure of the Java application we will build is shown in **Fig. 2.7** below.

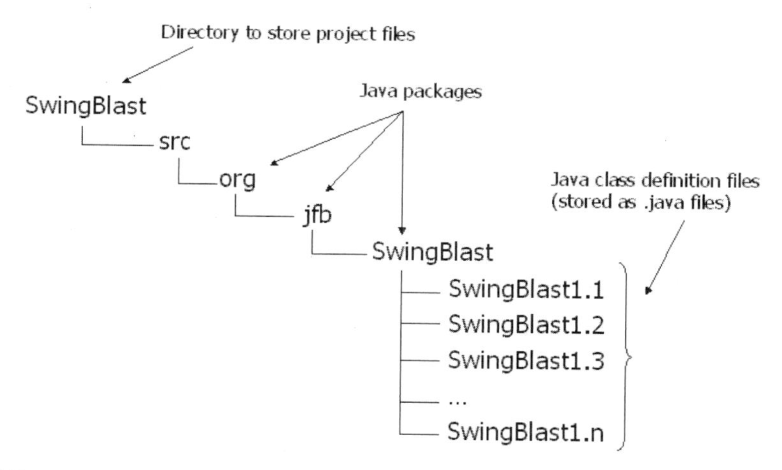

Fig. 2.7. Layout of the SwingBlast application

As depicted in **Fig. 2.7**, we define a project directory called `SwingBlast` to store the project files. We create a `src` (source) directory, in which we will create the packages `org`, `org.jfb` and `org.jfb.SwingBlast` to provide a default hierarchy for the class files. This layout also helps to group the necessary functionalities of the application, for example, by placing all the GUI classes in the SwingBlast package, all the source code files in the src directory and so on . `SwingBlast1.1`, ..., `1.n`, etc., are the Java class definition files, where the numbers refer to versions of the software as we build functionality step-by-step. For the SwingBlast application, the package name we will use in our Java class definition files will be `org.jfb.SwingBlast`. After the package is declared, we name any `import` statements to be included in the program. Import statements load the classes that encapsulate functions necessary for the application to run. Since classes are contained in packages for the purpose of grouping common functionalities together, entire packages may be imported, if necessary. By using wildcards with import statements for example,

```
import java.awt.*;
```

we can ensure that all classes in the AWT package, which provide the Java graphical user interface elements, are available to the application.

As we mentioned earlier, the `SwingBlast` application takes data input from the user and responds to the input by taking appropriate actions. To make the application respond appropriately to user initiated actions, we need to add what are known as *event listeners* to the code. This functionality allows us to add *events* to menu buttons that respond to simple actions such as clear user input or quit the application, etc., as well as complex functionality, some of which we will demonstrate in this Chapter. To begin with, we will learn the basics of the Java event model and see how to add events and event listeners in the next few sections.

Java Event Model

The Java Event Model is based on the *Observer design pattern* also known as the *Publish-Subscribe design pattern* and a delegation model that allows a source to propagate an event to the relevant observer. The *Publish-Subscribe design pattern* is based on the Observer pattern where the *Observer* object listens for events from the *Subject* object. The Publish-Subscribe design pattern is similar to the Observer design pattern except for additional element called the *Event Channel* that separates the Observer (called *Subscriber* in the Publish-Subscribe design pattern) and the Subject (called *Publisher* in the Publish-Subscribe design pattern). The Event Channel performs the role of a messaging hub to broadcast events from Publishers to all the associated Subscribers.

Java uses what are known as *EventListener* objects to listen to changes to AWT or Swing components. Under this model, observers can be registered to listen to an object via *Listener* methods depending on the type of the listener or the kinds of events one is interested in. The general format for such methods is `addXXXListener()`, for example, `addMouseListener(MouseListener l)`, which is a method to listen to any mouse event generated by the object the listener is registered to. The listener object provides a *callback* method that is called by the object that is generating the event. The callback method will have the appropriate parameters that define such data as the source (for example, JButton, JPanel, or a main window, etc.) and type of event (for example, a mouse click event, or a focus event when selecting a particular Swing component or an action event, like pressing a submit button).

In Java, all events are executed in the same thread as the window painting event (via `paint()`). This thread is called the *event-dispatching thread*. For this reason, code in an event listener should be fast to execute to avoid interference with the drawing events.

Two types of events are defined in Java: *low-level* events and *semantic* events. Low-level events represent system related events that emerge from objects such as mouse and keyboard, etc., while semantic events arise from operations such as clicking on a button, selecting a text in a drop down box, etc. Depending on the situation, it is advisable to listen to semantic events whenever possible since they are more specific in nature - for example, listening for a button event inside the component that contains the button, rather than a mouse event, which can occur outside of a component.

Adding Events to Applications

To add events to applications, we will need to add two import statements at the beginning of our code:

```
import java.awt.event.ActionEvent;

import java.awt.event.ActionListener;
```

These Java packages provide the classes that are needed for triggering and handling events. Let's take the example of making the SwingBlast application respond to actions initiated by the user by clicking on the `Quit` button under the `SwingBlast` Menu. To create the Quit button, we create an object called `quitItem` of type `JMenuItem` with the following piece of code:

```
quitItem = new JMenuItem("Quit");
```

To associate `quitItem` with a mouse click event that leads to closing the application, we first instantiate an `ActionListener`. Next we register the new listener to receive events from this button by calling the button's `addActionListener` method:

```
quitItem.addActionListener(new ActionListener() {
    public void actionPerformed(ActionEvent e) {
        System.exit(0);
    }
});
```

Actions triggered by mouse events such as a button click will also call the `actionPerformed` method from that listener and pass it an `ActionEvent` object as shown in the code above. That `ActionEvent` object contains all the properties of this event. In the early days, in C, you would have to catch the system interrupts and analyze the interrupt number received to figure out the type of event (viz., a keyboard or a mouse action or a USB port sending or receiving information, etc.). In Java, Swing does that for you by encapsulating all the hardware interactions into its event framework. This is undoubtedly much easier and means less work for the Java coder. Inside that `actionPerformed` method, all we need to do is to simply read the `ActionEvent` properties and code the appropriate action to respond to the event.

The code to handle events associated with the `Clear` button is constructed in a similar manner. The text box to enter sequences was earlier created as an object of type JTextArea using the code:

```
sequenceArea = new JTextArea();
```

The event handling code for the `Clear` button is similar, except that the exact action specified is that the text in the sequenceArea box is set to nothing (""):

```
clearButton.addActionListener(new ActionListener() {
    public void actionPerformed(ActionEvent e) {
        sequenceArea.setText("");
    }
});
```

Designing the SwingBlast GUI

We can now create the first version (1.1) of the `SwingBlast` application. `SwingBlast` version 1.1 will have a text box to enter sequence data, a `Clear` button to delete the entered sequence and a menu bar for quitting the application (**Fig. 2.8**).

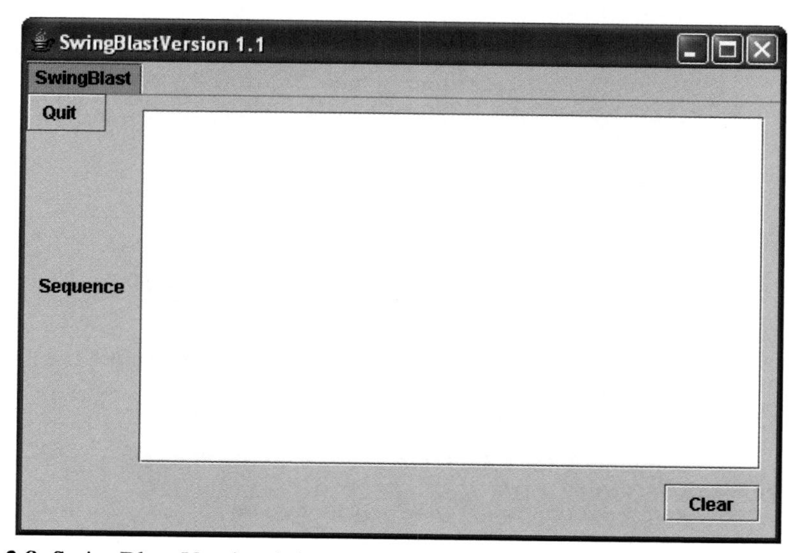

Fig. 2.8. SwingBlast Version 1.1

Let's now write the code that will create `SwingBlast` version 1.1. At the most basic level, our code will look like **Listing 2.1**.

Listing 2.1. Coding SwingBlast version 1.1

```
package org.jfb.SwingBlast;

import javax.swing.*;
import java.awt.*;

public class SwingBlast1_1 extends JFrame {
    private static final String APP_NAME = "SwingBlast";
    private static final String APP_VERSION = "Version
1.1";
    private static final Dimension APP_WINDOW_SIZE = new
Dimension(500, 300);

    private JComponent newContentPane;
    private JTextArea sequenceArea;
    private JScrollPane scrollPaneArea;
```

```
    private JButton clearButton;
    private JMenuItem quitItem;

    public SwingBlast1_1() {
        super(APP_NAME + APP_VERSION);
        setDefaultCloseOperation(JFrame.EXIT_ON_CLOSE);
        newContentPane = new JPanel();
        newContentPane.setLayout(new BorderLayout());
        setContentPane(newContentPane);

        JMenuBar menu = new JMenuBar();
        JMenu swingBlastMenu = new JMenu(APP_NAME);
        quitItem = new JMenuItem("Quit");
        swingBlastMenu.add(quitItem);
        menu.add(swingBlastMenu);
        setJMenuBar(menu);

        // The sequence pane
        JPanel sequencePanel = new JPanel();
        JLabel sequence = new JLabel("Sequence");
        sequenceArea = new JTextArea();
        sequenceArea.setLineWrap(true);
        scrollPaneArea = new JScrollPane(sequenceArea);
        sequencePanel.setLayout(new
BoxLayout(sequencePanel, BoxLayout.LINE_AXIS));
        sequencePanel.add(sequence);
        sequencePanel.add(Box.createRigidArea(new
Dimension(10, 0)));
        sequencePanel.add(scrollPaneArea);

sequencePanel.setBorder(BorderFactory.createEmptyBorder(10,
0, 10, 0));

        //Lay out the buttons from left to right
JPanel buttonPane = new JPanel();
        clearButton = new JButton("Clear");

        buttonPane.setLayout(new BoxLayout(buttonPane,
BoxLayout.LINE_AXIS));
        buttonPane.add(Box.createHorizontalGlue());
        buttonPane.add(Box.createRigidArea(new
Dimension(10, 0)));
        buttonPane.add(clearButton);

        JPanel jPanel = new JPanel();
        jPanel.setLayout(new BorderLayout());
        jPanel.setBorder(BorderFactory.createEmptyBorder(0,
10, 10, 10));
        jPanel.add(sequencePanel, BorderLayout.CENTER);
        jPanel.add(buttonPane, BorderLayout.SOUTH);

        newContentPane.add(jPanel, BorderLayout.CENTER);
        newContentPane.setPreferredSize(APP_WINDOW_SIZE);
```

```
            //Display the window
            pack();
            Dimension                    screenSize           =
Toolkit.getDefaultToolkit().getScreenSize();
            setLocation((screenSize.width                        -
APP_WINDOW_SIZE.width) / 2,
                        (screenSize.height                       -
APP_WINDOW_SIZE.height) / 2);
            setVisible(true);

        }

        public static void main(String[] args) {
            SwingUtilities.invokeLater(new Runnable() {
                public void run() {
                    final     SwingBlast1_1     view     =     new
SwingBlast1_1();
                }
            });
        }
    }
```

As described earlier, we begin by declaring a package, which in this case is named after the `SwingBlast` application that we are building. The common prefix `jfb` is short for Java for Bioinformatics. Since we are creating a Swing based GUI to manage sequence input and analysis, we have named the class "SwingBlast". The suffix 1_1 at the end of the class name reflects the fact that this is version 1.1 of the `SwingBlast` application.

A simplified general format of the class declaration is as follows:

```
class_modifiers class <class_name> extends <superclass_name>
        {
                /* list of class data fields */
                /* list of class methods     */
        }
```

In our case, the modifier for the `SwingBlast1_1` class is *public* which means other methods or classes outside of this class may access this class:

```
  public class SwingBlast1_1 extends JFrame {
    ...
  }
```

By convention, there can be only one *public* class in a Java file; further, the name of the Java file must match the name of the *public* class. For this reason, the code in **Listing 2.1** must be stored in a file called `SwingBlast1_1.java`.

Note the use of the `extends` keyword in the class declaration. The `extends` keyword indicates that the `SwingBlast1_1` class inherits methods from the class *JFrame*. In object oriented terminology, `SwingBlast1_1` is called the *sub* or *child* class while *JFrame* which it derives from is called the *parent* (or *super*) class. The `extends` keyword obviates the need for instantiating *JFrame* separately in the `SwingBlast1_1` class to access its methods. Inside the `SwingBlast1_1` class, we can call any of the methods available in the parent *JFrame* class.

JFrame is a *Swing container* that serves as the top-level or main application window. Top-level Swing containers provide space within which other Swing components can position and draw themselves. Swing components are also called "*lightweight components*" because they are written in Java versus *AWT* components or "*heavyweight components*" which are native components (written in C or C++, etc.) wrapped into Java classes. It is important to know what class of components are being used. As a rule of thumb, Swing and AWT components should not be mixed or used together in the same application, as this may lead to unpredictable behavior during repainting, and may make the application hard to debug.

The Swing framework provides a mechanism for interactions with individual components through *event handling*. This is what the two import statements at the top of our code in **Listing 2.1** do:

```
import javax.swing.*;

import java.awt.*;
```

The first package provides a set of lightweight components while the second contains all classes for dealing with graphics, events, images, etc. **Fig. 2.9** shows the superclass hierarchy of the JFrame class where each subclass is shown below its parent class. According to this scheme, the *JFrame* class is derived from the *Frame* class, which in turn is derived from the *Window* class and so on. The *Frame* class defines a top-level window with a title and a border and has methods such as `getTitle`, `setTitle` etc., which respectively get and set the title of the frame. By definition, the JFrame class derives these methods from the *Frame* class (and other superclasses). Every main window that contains Swing components should be implemented with a JFrame. Examples of other containers are *JApplet*, *JWindow* and *JDialog*. In our application, JFrame will serve as the top-level container. JFrame in turn will provide the framework to contain other components like for example *JPanel*, *JButton* and *JMenu*, etc.

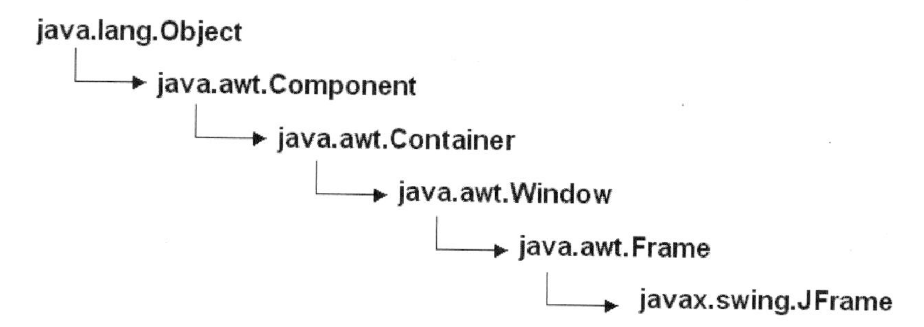

Fig. 2.9. Class hierarchy of the JFrame class

The next three lines of code define constants for setting the name (`SwingBlast`), version (1.1) and the window size (500 x 300 pixels) of the application. We will use upper case names separated with underscores '_', as a naming convention for our constants :

```
private static final String APP_NAME = "SwingBlast";
```

private limits the accessibility of the variable called `APP_NAME` to other objects within the same class. The keyword *Static* means that the value of the variable is shared by any object of that same class (this also defines what is known as the class variable). This means that if one object modifies it, the other object can see the new value. A non-static variable, on the other hand, is modifiable only by the object instantiated from within the same class. The keyword *final* means that the variable cannot be changed and therefore it is a constant. The constants `APP_NAME` and `APP_VERSION` are of type *String* as indicated in the code. To summarize, `APP_NAME` is a constant accessible only from within the class and it has the same value for any object belonging to this class.

The next 5 lines declare Swing components of the types *JComponent*, *JTextArea*, *JScrollPane*, *JButton* and *JMenuItem* respectively. All Swing components (except top-level containers) whose names begin with "J" are derived from and inherit from the *JComponent* class such as *JTextArea*, *JPanel*, *JScrollPane*, *JButton*, and *JmenuItem*. *JComponent* is thus the base class for all these Swing components.

The next line:

```
public SwingBlast1_1() {
```

defines the constructor for the `SwingBlast1_1` class. Note that it is declared *public*, has the same name as the class itself and does not return anything. The `SwingBlast1_1` constructor also does not accept any parameters and therefore is the default constructor for the `SwingBlast1_1` class.

The *super* keyword in the `SwingBlast1_1` constructor calls the constructor of the superclass (hence the use of the term "*super*") - which in this case is JFrame, since `SwingBlast1_1` "extends" JFrame. Next it passes the *String* variables `APP_NAME` and `APP_VERSION` to the JFrame constructor to set the name and version of the application. The description of the JFrame constructor that is used is shown below. This information is available from the Java 2 API documentation (**Fig. 2.10**).

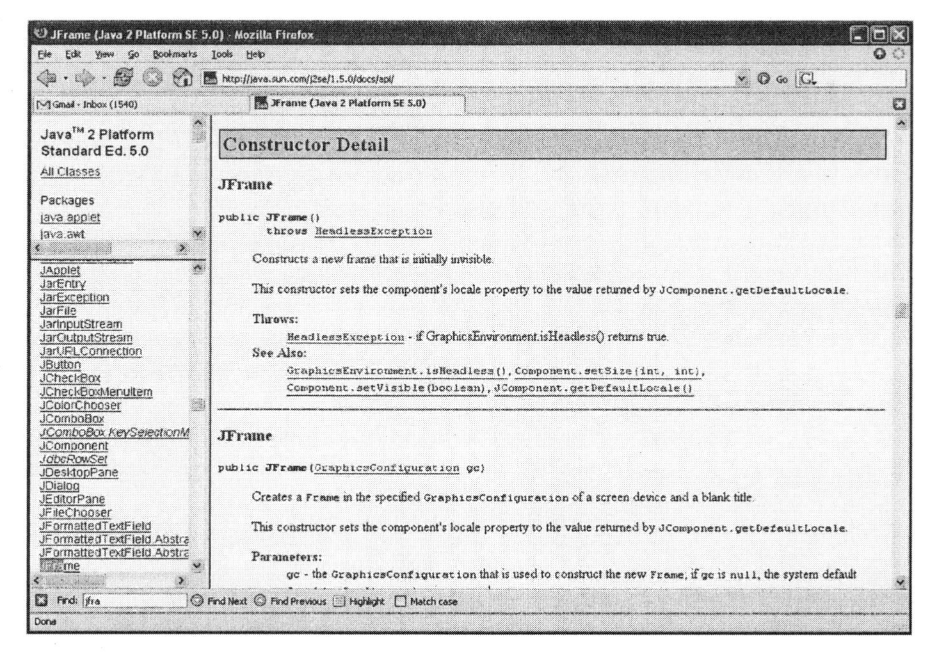

Fig. 2.10. Java 2 API documentation on JFrame

```
Name: JFrame(String title)
Description: Creates a new, initially invisible Frame with
the specified title.
```

The same result can also be achieved by explicitly setting the title as follows:

```
setTitle(APP_NAME + " " + APP_VERSION);
```

The line:

```
setDefaultCloseOperation(JFrame.EXIT_ON_CLOSE);
```

uses the `setDefaultCloseOperation` method from the `JFrame` class which is defined as follows:

```
public void setDefaultCloseOperation(int operation)
```

and sets the operation that we want the application to perform by default when the user attempts to close the frame. We have specified the operation to exit the application by using the `EXIT_ON_CLOSE` option.

Coding the SwingBlast GUI

The general scheme for creating and adding Swing components to an application consists of the following steps:

1. Create an instance of a top-level container such as Jframe
2. Use a *layout manager* to specify the location and size of the components
3. Specify the top-level container's content pane to hold the individual GUI elements

To begin with, we create an instance of the JPanel class (called `newContentPane`), which defines a generic container as the top-level container. We will use this container to hold our GUI elements. Components are positioned inside a top-level container using what are known as *layout managers* in Java. The area within a top-level container where individual components (labels, buttons, etc.) are placed is called the *content pane*. To specify the *content pane* of the `newContentPane` component as the *content pane* for storing the visible elements of the

`SwingBlast` application, we use the top-level container's `setContentPane` method:

```
newContentPane = new JPanel();
newContentPane.setLayout(new BorderLayout());
  setContentPane(newContentPane);
```

Here we have used the *BorderLayout layout manager* to align and position the components. Next we add the menu bar (called `SwingBlast`) and a single menu item ("`Quit`"):

```
JMenuBar menu = new JMenuBar();
JMenu swingBlastMenu = new JMenu(APP_NAME);
quitItem = new JMenuItem("Quit");
swingBlastMenu.add(quitItem);
menu.add(swingBlastMenu);
setJMenuBar(menu);
```

Note that components are added using the `add` method as shown here for the `SwingBlast` menu:

```
menu.add(swingBlastMenu);
```

Next we create the sequence pane and add a component called `sequenceArea` of the type *JTextArea* that simply defines an area for entering text:

```
            // The sequence pane
            JPanel sequencePanel = new JPanel();
            JLabel sequence = new JLabel("Sequence");
            sequenceArea = new JTextArea();
            sequenceArea.setLineWrap(true);
            scrollPaneArea = new JScrollPane(sequenceArea);
            sequencePanel.setLayout(new
BoxLayout(sequencePanel, BoxLayout.LINE_AXIS));
            sequencePanel.add(sequence);
            sequencePanel.add(Box.createRigidArea(new
Dimension(10, 0)));
            sequencePanel.add(scrollPaneArea);

sequencePanel.setBorder(BorderFactory.createEmptyBorder(10,
0, 10, 0));
```

To provide scrolling capabilities inside the text area (especially for large sequences), we have associated the *JScrollPane* object with the `sequenceArea`. The `clear` button is added in a similar fashion. Finally, we add the *main()* method to the program. The *main()* method actually performs the job of creating an instance of the class and running the application. The *Java Virtual Machine (JVM)* calls this `main()` method when we pass the class name to it. Every Java application must contain a *main()* method whose signature looks like this:

```
public static void main(String[] args) {
      // statements;
}
```

The JVM would eventually complain about a class if the `main()` method was missing. The simplified general format for a method in Java is:

```
method_modifier return_type method_name (arguments) {
     body of the method;
}
```

In our case, the method looks like this:

```
public static void main(String[] args) {
  SwingUtilities.invokeLater(new Runnable() {
    public void run() {
      final SwingBlast1_1 view = new SwingBlast1_1();
    }
  });
 }
}
```

The line:

```
SwingUtilities.invokeLater(new Runnable() { }
```

indicates that the painting of the GUI takes place in a separate *thread* (the *AWT thread* or *the event-dispatching thread*) and is a way of separating the GUI processes from the business processes (such as a BLAST operation) as strongly advised in the Java guidelines.

> A *thread* is a process that is capable of running concurrently alongside other *threads* or processes.
>
> The *event-dispatching thread* is the thread responsible for handling events and repainting of components. It is therefore very important to avoid any running heavy resource consuming code in the *event-dispatching thread.*

The keyword *Runnable* defines the type of object that will run in a new thread. The *invokeLater()* method causes the event-dispatching thread to call the `run()` method of the Runnable object which is passed to *invokeLater()* method after all pending events (such as repainting a component, etc.) are processed. The `run()` method of the Runnable object is in charge for creating a SwingBlast object through the constructor method of SwingBlast, which in turn performs all the specified actions, such as creating the top level window, setting its name and laying out the GUI elements, etc.

Compile and run the code shown in **Listing 2.1**. As you will notice, the basic framework as described above does not do anything useful apart from displaying the graphical interface as shown in **Fig. 2.8**. The only events the application can respond to so far are the default Minimize, Maximize and Close operations through icons located on the top right of the application window.

Coding the SwingBlast Business Logic

We will begin the process of building the business logic into the application by adding code that will format the user entered sequence into the commonly used Fasta format. We will simultaneously add code that will calculate and display the size of the input sequence. We will then incorporate a simple algorithm to determine the sequence type – that is, if the user entered sequence is nucleotide or protein.

The Fasta format as defined earlier contains a header that begins with the greater than symbol (>) and contains information about the sequence such as sequence identifiers and size, etc. (which may be delimited by separators such as vertical bars or spaces) on the first line and is followed on the second line with the actual sequence (**Fig. 2.6**).

So how do we get the sequence entered in the text area to rearrange itself in the Fasta format? As with any programming language there are more than one ways of achieving this. We will use a method based on *Focus events* to implement this. *Focus events* are triggered whenever a component such as text area gains or loses focus. *Focus events* associated with a particular component can be obtained by registering a *FocusListener* with the component. When the component gains or loses focus, the relevant method in the *listener* object (*focusGained* or *focusLost*, respectively) is invoked, and the *FocusEvent* is passed to it. The general method to do this is shown in **Listing 2.2**.

Listing 2.2. Adding Focus events and listeners to SwingBlast

```
sequenceArea.addFocusListener(new FocusListener() {
  public void focusGained(FocusEvent e) {

  }

  public void focusLost(FocusEvent e) {
    // add statements here
  }
});
```

We will design the code such that after a sequence has been added to the text area, it will be converted into the Fasta format as soon as the text area loses focus (for example, when a user navigates away from the text area to another part of the application). Conversely, no action will be performed when the `sequenceArea` component gains focus. We therefore want to add program logic in the *focusLost* method, which gets activated after a component loses focus, to achieve this. **Listing 2.3** shows how to implement this.

Listing 2.3. Programming the `focusLost` method

```
sequenceArea.addFocusListener(new FocusListener() {
  public void focusGained(FocusEvent e) {

  }

  public void focusLost(FocusEvent e) {
    // Retrieve the sequence in the text area
    String seqText = sequenceArea.getText();

    // Convert the sequence into Fasta format
    String header = null;
    int seqLength = 0;
    String sequence = "";
    String fastaSeq = "";

    seqText = seqText.replaceAll("\\s", "");
    sequence = seqText.toLowerCase();
    header = "> Sequence1";
    seqLength = seqText.length();
    fastaSeq = header + "|" + seqLength + "\n" +
sequence;
```

```
            sequenceArea.setText(fastaSeq);
        }

    });
```

For the header part of the Fasta sequence, we will add a generic label (called "sequence1") to represent the name of the raw sequence entered by the user followed by a vertical bar and the size of the sequence for the purpose of illustration. Plug this into the main code and test the application by pasting a sequence (such as the first few hundred bases of the CFTR gene sequence shown below) into it.

```
AATTGGAAGCAAATGACATCACAGCAGGTCAGAGAAAAAGGGTTGAGCGGCAGGCACCCAG
AGTAGTAGGTCTTTGGCATTAGGAGCTTGAGCCCAGACGGCCCTAGCAGGGACCCCAGCGC
CCGAGAGACCATGCAGAGGTCGCCTCTGGAAAAGGCCAGCGTTGTCTCCAAACTTTTTTTC
AGCTGGACCAGACCAATTTTGAGGAAAGGATACAGACAGCGCCTGGAATTGTCAGACATAT
ACCAAATCCCTTCTGTTGATTCTGCTGACAATCTATCTGAAAAATTGGAAAGAGAATGGGA
TAGAGAGCTGGCTTCAAAGAAAAATCCTAAACTCATTAATGCCCTTCGGCGATGTTTTTTC
TGGAGATTTATGTTCTATGGAATCTTTTTATATTTAGGGGAAGTCACCAAAGCA
```

You will see that once the text area loses focus, for example, by clicking on the `SwingBlast` menu, the sequence is converted into lower case and formatted into the Fasta format (**Fig. 2.11** and **Fig. 2.12**).

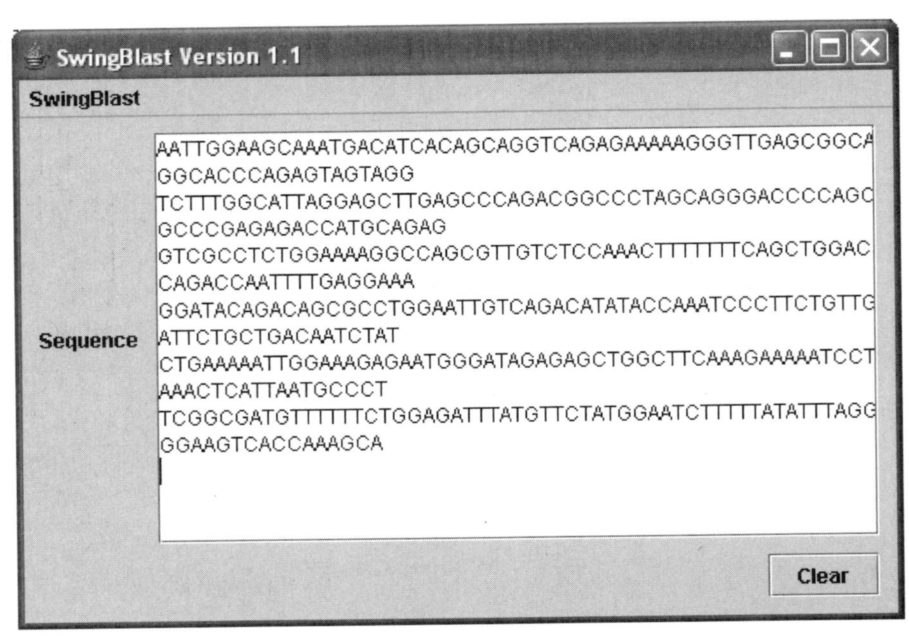

Fig. 2.11. Unformatted nucleotide sequence

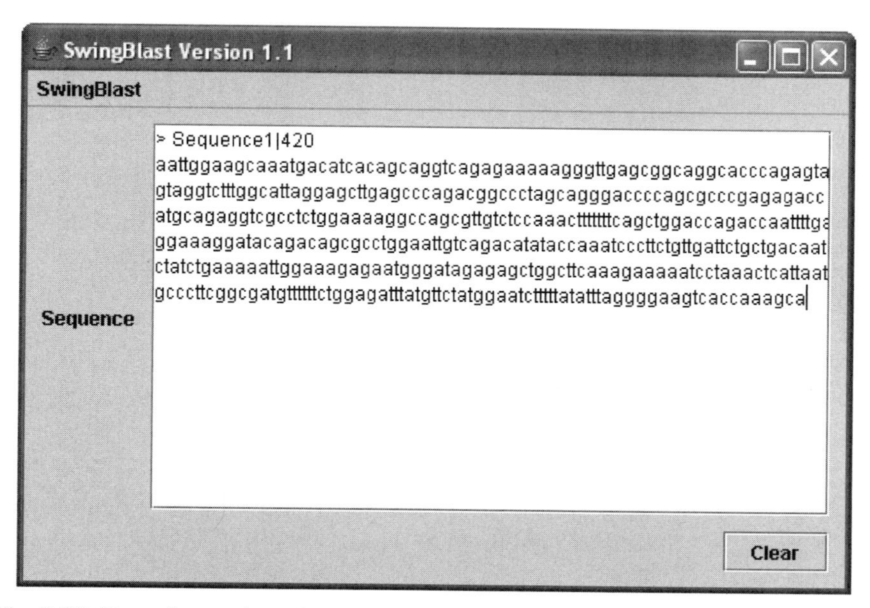

Fig. 2.12. Fasta formatting of sequences (Text area loses focus)

In addition, a header line is added as specified in the code along with the length of the sequence. Although the logic to convert raw sequence into Fasta format does work as described, we need to incorporate a way to tell the *FocusEvent* method not to take any action if the sequence is already in the Fasta format (either because the sequence was pasted in the Fasta format or because it was formatted by the user formatted by the user using the FocusLost method) and therefore does not need formatting. This is easily done by checking for the presence of the ">" character at the beginning of the sequence as shown in **Listing 2.4** below.

Listing 2.4. Checking for Fasta formatting of sequences

```
sequenceArea.addFocusListener(new FocusListener() {
  public void focusGained(FocusEvent e) {

 }

 public void focusLost(FocusEvent e) {
   // Retrieve the sequence in the text area
   String seqText = sequenceArea.getText();

   int idx = seqText.indexOf(">");
   boolean fastaFormatted = idx != -1;
```

```
            String header = null;
            int seqLength = 0;
            String sequence = "";
            String fastaSeq = "";
    // Check if sequence is in Fasta format
            if (fastaFormatted) {
              int returnIdx = seqText.indexOf("\n");
              header = seqText.substring(0, returnIdx);
              fastaSeq = seqText.substring(returnIdx + 1,
    seqText.length()).replaceAll("\\s", "").toLowerCase();
              fastaSeq = seqText;
            } else {
              seqText = seqText.replaceAll("\\s", "");
              fastaSeq = seqText.toLowerCase();
              header = "> Sequence1";
              seqLength = seqText.length();
            }

    // Convert the sequence into Fasta format if not Fasta
//formatted
            if (!fastaFormatted) {
              fastaSeq = header + "|" + seqLength + "\n" +
fastaSeq;

            }
            sequenceArea.setText(fastaSeq);
          }
```

To make the sequence align properly, we will use a monospace font such as Courier. The code to do this is as follows:

```
final Font sf = sequenceArea.getFont();
Font f = new Font("Monospaced", sf.getStyle(), sf.getSize());
sequenceArea.setFont(f);
```

Run the code again. This time the sequence is properly aligned (**Fig. 2.13**).

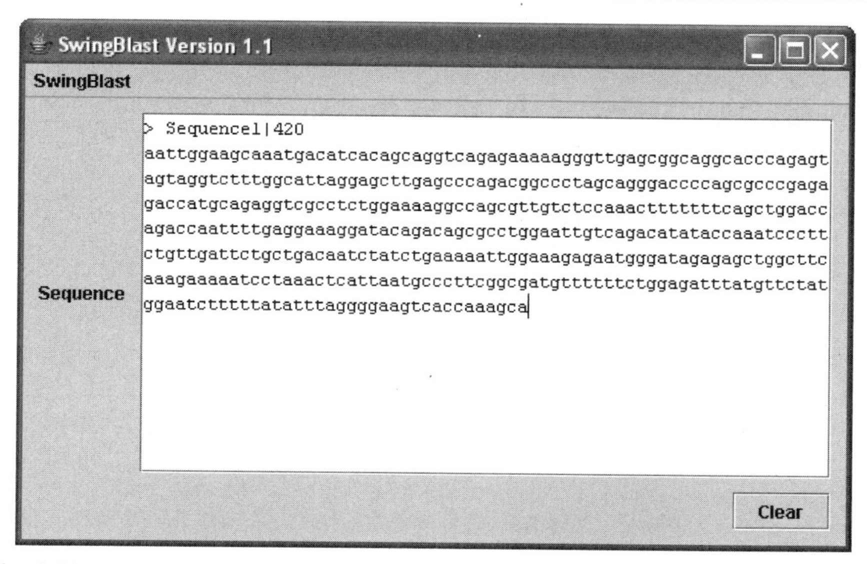

Fig. 2.13. Using monospace font to format sequences

Determining Sequence Type: Nucleotide or Protein?

Now that we have formatted the sequence and calculated its size, lets plug in functionality into the SwingBlast application that will determine if the entered sequence is nucleotide (DNA or RNA) or protein. Note that RNA, like DNA is a polymer composed of four nucleotides. The difference between RNA and DNA is the nature of the sugar moiety: RNA has the ribose sugar, while DNA has the deoxyribose sugar. RNA has the same purine bases as DNA: adenine (A) and guanine (G) and the same pyrimidine cytosine (C), but instead of thymine (T), it uses the pyrimidine uracil (U).

Determination of sequence type is done with an algorithm that takes into account information on the natural composition of nucleotide and protein sequences. According to the algorithm, if:

1. Total number of nculeotides (that is, sum of A, T, G and C's) divided by the total length of the sequence is greater that 0.85, it is a DNA sequence

2. Total number of A, T, G, C and U's divided by the total length of the sequence is greater that 0.85, it is an RNA sequence

If neither of these two conditions is met, the sequence is assumed to be a protein sequence. Note that we are not using the extended DNA/RNA alphabet that includes symbols for sequence ambiguity as defined in the International Union of Pure and Applied Chemistry (IUPAC) and International Union of Biochemistry (IUB) nucleotide and amino acid nomenclature. Instead, we are assuming the DNA alphabet to be composed of the four bases A (adenine), T (thymine), G (guanine), C (cytosine) and N, the RNA alphabet to be composed of A (adenine), U (uridine), G (guanine), C (cytosine) and N (where N is any nucleotide base) and the amino acid alphabet to be composed of A (alanine), C (cysteine), D (aspartate), E (glutamic acid), F (phenylalanine), G (glycine), H (histidine), I (isoleucine), K (lysine), L (leucine), M (methionine), N (asparagine), P (proline), Q (glutamine), R (arginine), S (serine), T (threonine), V (valine), W (tryptophan) and Y (tyrosine).

Let's see how this algorithm works with an example. Take the partial mRNA sequence of the human CFTR gene (gi: 90421312) as shown below:

```
AAUUGGAAGCAAAUGACAUCACAGCAGGUCAGAGAAAAAGGGUUGAGCGGCAGGCACCCAG
AGUAGUAGGUCUUUGGCAUUAGGAGCUUGAGCCCAGACGGCCCUAGCAGGGACCCCAGCGC
CCGAGAGACCAUGCAGAGGUCGCCUCUGGAAAAGGCCAGCGUUGUCUCCAAACUUUUUUUC
AGCUGGACCAGACCAAUUUUGAGGAAAGGAUACAGACAGCGCCUGGAAUUGUCAGACAUAU
ACCAAAUCCCUUCUGUUGAUUCUGCUGACAAUCUAUCUGAAAAAUUGGAAAGAGAAUGGGA
UAGAGAGCUGGCUUCAAAGAAAAAUCCUAAACUCAUUAAUGCCCUUCGGCGAUGUUUUUUC
UGGAGAUUUAUGUUCUAUGGAAUCUUUUUAUAUUUAGGGGAAGUCACCAAAGCAGUACAGC
CUCUCUUACUGGGAAGAAUCAUAGCUUCCUAUGACCCGGAUAACAAGGAGGAACGCUCUAU
CGCGAUUUAUCUAGGCAUAGGCUUAUGCCUUCUCUUUAUUGUGAGGACACUGCUCCUACAC
CCAGCCAUUUUUGGCCUUCAUCACAUUGGAAUGCAGAUGAGAAUAGCUAUGUUUAGUUUGA
UUUAUAAGAAGACUUUAAAGCUGUCAAGCCGUGUUCUAGAUAAAAUAAGUAUUGGACAACU
UGUUAGUCUCCUUUCCAACAACCUGAACAAAUUUGAUGAAGGACUUGCAUUGGCACAUUUC
GUGUGGAUCGCUCCUUUGCAAGUGGCACUCCUCAUGGGGCUAAUCUGGGAGUUGUUACAGG
CGUCUGCCUUCUGUGGACUUGGUUUCCUGAUAGUCCUUGCCCUUUUU
```

We will call this sequence with a size of 840 bases "S1". Lets start by removing all A, T, G and C's from the sequence. The length of the sequence without A, T, G and C's is 237; lets call this sequence S2.

Number of A, T, G and C's in the sequence = S1 – S2 = 603. Next we remove all the U's from the sequence that remain after removing the A, T, G and C's (that is, the sequence S2). The length of the sequence after removing all the U's is zero (since all we had left were U's). Lets call this S3. The total number of U's in the sequence is therefore S2 – S3 is 237.

Now let's calculate the relative proportions of DNA and RNA alphabets in the sequence.

$$(A + T + G + C)/\text{Total} = 603/840 = 0.72$$

According to the algorithm, since this is less than 0.85, it cannot be a DNA sequence.

$$(A + T + G + C + U)/\text{Total} = (603 + 237)/840 = 1$$

Since this is > 0.85, this is an RNA sequence. We can now write the code using the above reasoning. Since we will use *regular expression* matching to parse the sequence, we will first import the appropriate libraries to do so:

```
import org.apache.regexp.RE;

import org.apache.regexp.RESyntaxException;
```

We declare the magic 0.85 number as a threshold:

```
private static final double SEQ_THRESHOLD = 0.85;
```

The `getSequenceType()` method that implements the algorithm is as follows:

```
    public static int getSequenceType(String sequence) throws
RESyntaxException {
        RE re = new RE("[actgnACGTN]+");
        String[] strings = re.split(sequence);
        int numbOfLettersOtherThanATGCNs = 0;

        for (int i = 0; i < strings.length; i++) {
          numbOfLettersOtherThanATGCNs                           +=
strings[i].length();
        }
        int length = sequence.length();
        int          numbOfACGTNs          =          length          -
numbOfLettersOtherThanATGCNs;

        re = new RE("[uU]+");
        strings = re.split(sequence);
        int numbOfLettersOtherThanUs = 0;

        for (int i = 0; i < strings.length; i++) {
          numbOfLettersOtherThanUs += strings[i].length();
        }
```

```
        int      numbOfUs      =      sequence.length()     -
numbOfLettersOtherThanUs;

        if (numbOfACGTNs / (double) length > SEQ_THRESHOLD) {
          return TYPE_DNA;
        } else if ((numbOfACGTNs + numbOfUs) / (double)
length > SEQ_THRESHOLD) {
          return TYPE_RNA;
        } else {
          return TYPE_PROTEIN;
        }
      }
```

With this code in place, we get the following results for the partial sequences of the human CFTR nucleotide (**Fig. 2.14** and **Fig. 2.15**) and protein (**Fig. 2.16** and **Fig. 2.17**).

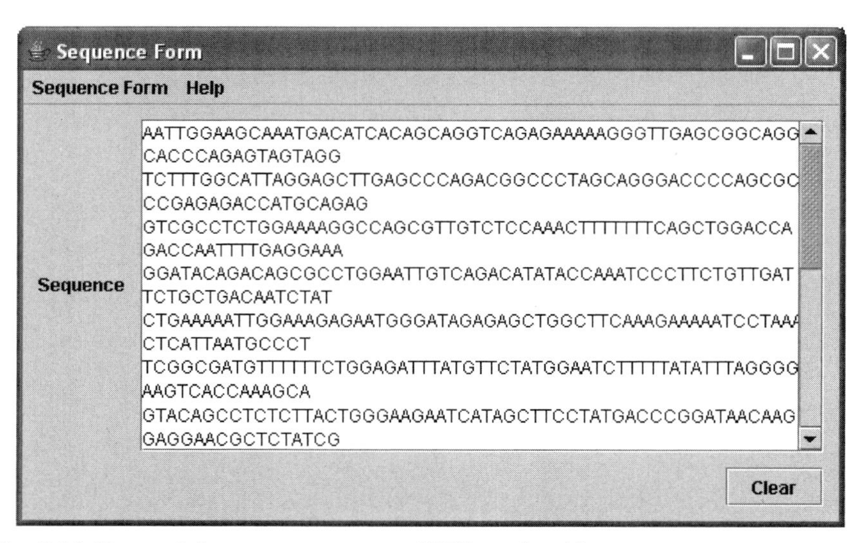

Fig. 2.14. Determining sequence type - CFTR nucleotide sequence

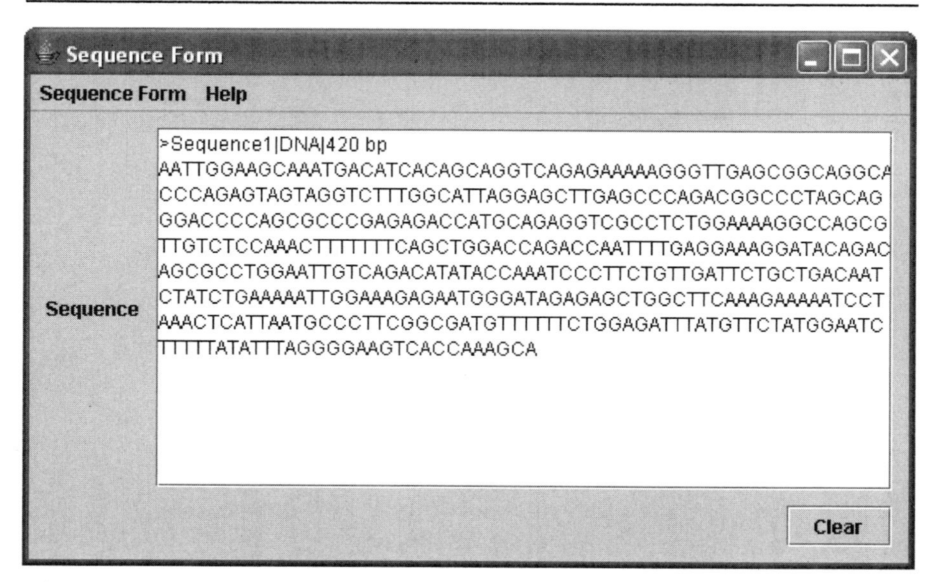

Fig. 2.15. Determining sequence type - CFTR nucleotide sequence

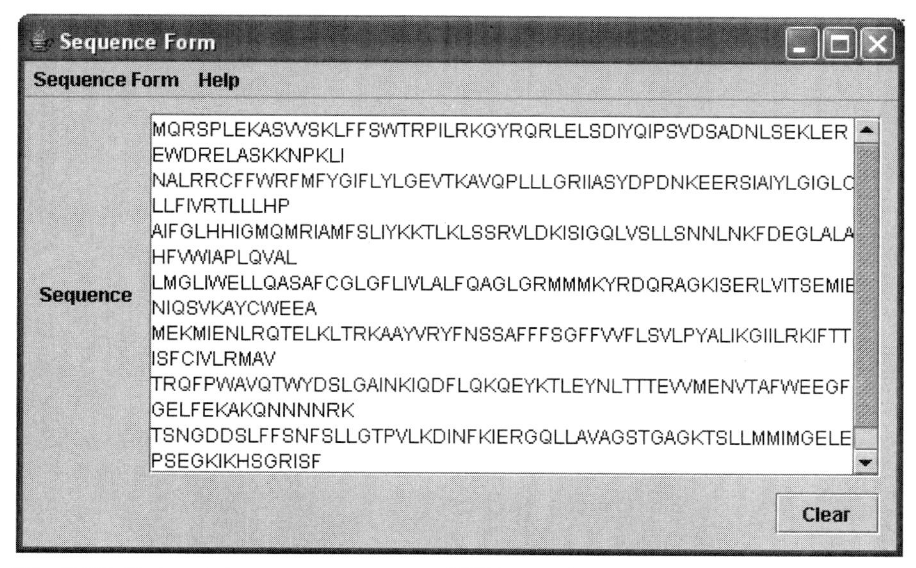

Fig. 2.16. Determining sequence type - CFTR protein sequence

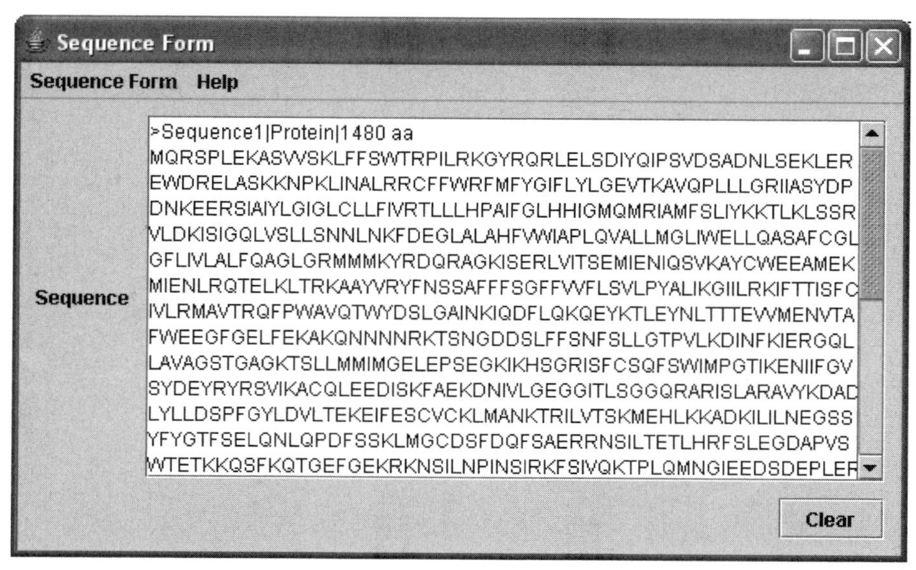

Fig. 2.17. Determining sequence type: CFTR protein sequence

We will call this `SwingBlast` version 1.2. The complete code is described in **Listing 2.5**.

Listing 2.5. Determining sequence type

```
package org.jfb.SwingBlast;

import org.apache.regexp.RE;
import org.apache.regexp.RESyntaxException;

import javax.swing.*;
import java.awt.*;
import java.awt.event.ActionEvent;
import java.awt.event.ActionListener;
import java.awt.event.FocusEvent;
import java.awt.event.FocusListener;

public class SwingBlast1_2 extends JFrame {
  private static final String APP_NAME = "Sequence Form";
  private static final String APP_VERSION = "Version 1_2";

  private static final Dimension APP_WINDOW_SIZE = new
Dimension(450, 350);

  private static final int TYPE_DNA = 0;
  private static final int TYPE_RNA = 1;
  private static final int TYPE_PROTEIN = 2;

  private JComponent newContentPane;
```

```
    private JTextArea sequenceArea;
    private JScrollPane scrollPaneArea;
    private JButton clear;

    private JMenuItem aboutItem;
    private JMenuItem quitItem;
    private static final double SEQ_THRESHOLD = 0.85;

    public SwingBlast1_2() {
      super();
      seqFormInit();
    }

    private void seqFormInit() {
      setTitle(APP_NAME);
      setDefaultCloseOperation(JFrame.EXIT_ON_CLOSE);
      newContentPane = new JPanel();
      newContentPane.setOpaque(true);
      newContentPane.setLayout(new BorderLayout());

      setContentPane(newContentPane);

      // Create the menu bar
      JMenuBar menu = new JMenuBar();
      JMenu swingBlastMenu = new JMenu(APP_NAME);
      quitItem = new JMenuItem("Quit");
      swingBlastMenu.add(quitItem);
      menu.add(swingBlastMenu);

      JMenu helpMenu = new JMenu("Help");
      aboutItem = new JMenuItem("About");
      helpMenu.add(aboutItem);
      menu.add(helpMenu);
      setJMenuBar(menu);

      // Create the sequence pane
      JPanel sequencePanel = new JPanel();
      JLabel sequence = new JLabel("Sequence");
      sequenceArea = new JTextArea();
      Font font = sequence.getFont();
      sequenceArea.setFont(new  Font("Courier",  Font.PLAIN,
font.getSize()));
      sequenceArea.setLineWrap(true);
      scrollPaneArea = new JScrollPane(sequenceArea);

      sequencePanel.setLayout(new    BoxLayout(sequencePanel,
BoxLayout.LINE_AXIS));
      sequencePanel.add(sequence);
      sequencePanel.add(Box.createRigidArea(new Dimension(10,
0)));
      sequencePanel.add(scrollPaneArea);

sequencePanel.setBorder(BorderFactory.createEmptyBorder(10,
```

```
0, 10, 0));

      // Lay out the buttons from left to right
      JPanel buttonPane = new JPanel();
      clear = new JButton("Clear");

      buttonPane.setLayout(new          BoxLayout(buttonPane,
BoxLayout.LINE_AXIS));
      buttonPane.add(Box.createHorizontalGlue());
      buttonPane.add(Box.createRigidArea(new    Dimension(10,
0)));
      buttonPane.add(clear);

      JPanel jPanel = new JPanel();
      jPanel.setLayout(new BorderLayout());
      jPanel.setBorder(BorderFactory.createEmptyBorder(0, 10,
10, 10));
      jPanel.add(sequencePanel, BorderLayout.CENTER);
      jPanel.add(buttonPane, BorderLayout.SOUTH);

      newContentPane.add(jPanel, BorderLayout.CENTER);
      newContentPane.setPreferredSize(APP_WINDOW_SIZE);

      // Display the window
      pack();
      Dimension                  screenSize                 =
Toolkit.getDefaultToolkit().getScreenSize();
      setLocation((screenSize.width - APP_WINDOW_SIZE.width)
/ 2,
        (screenSize.height - APP_WINDOW_SIZE.height) / 2);
      setVisible(true);

      addListeners();
    }

    private void addListeners() {
      quitItem.addActionListener(new ActionListener() {
        public void actionPerformed(ActionEvent e) {
          System.exit(0);
        }
      });

      aboutItem.addActionListener(new ActionListener() {
        public void actionPerformed(ActionEvent e) {
          JOptionPane.showMessageDialog(SwingBlast1_2.this,
APP_NAME + " " + APP_VERSION,
            "About           "          +          APP_NAME,
JOptionPane.INFORMATION_MESSAGE);
        }
      });

      clear.addActionListener(new ActionListener() {
        public void actionPerformed(ActionEvent e) {
```

```
          sequenceArea.setText("");
        }
      });

      sequenceArea.addFocusListener(new FocusListener() {
        public void focusGained(FocusEvent e) {

        }

        public void focusLost(FocusEvent e) {
          // Check if the sequence is DNA, RNA or protein

          String text = sequenceArea.getText();

  // Format the sequence in FASTA format and retrieve the
  // sequence the user entered
          int idx = text.indexOf(">");
          boolean fastaFormatted = idx != -1;
          String seqText = null;
          String header = null;
          int seqLength = 0;
          String sequence = "";

          if (fastaFormatted) {
            int returnIdx = text.indexOf("\n");
            header = text.substring(0, returnIdx);
            sequence   =   text.substring(returnIdx   +   1,
text.length()).replaceAll("\\s", "").toLowerCase();
            seqText = text;
          } else {
            text = text.replaceAll("\\s", "");
            sequence = text.toLowerCase();
            header = ">Sequence1|";
            seqLength = text.length();
          }

          // Determine the sequence type
          int typeOfSequence = -1;
          try {
            typeOfSequence = getSequenceType(sequence);
          } catch (RESyntaxException e1) {
            e1.printStackTrace();
          }

          String type = null;
          String unitOfLength = null;

          switch (typeOfSequence) {
            case TYPE_DNA:
              type = "DNA";
              unitOfLength = " bp";
              break;
            case TYPE_RNA:
```

```
              type = "RNA";
              unitOfLength = " bp";
              break;
            case TYPE_PROTEIN:
              type = "Protein";
              unitOfLength = " aa";
              break;
            default:
              type = "N/A";
              unitOfLength = " N/A";
          }

          if (!fastaFormatted) {
            seqText = header + type + "|" + seqLength +
unitOfLength + "\n" + sequence.toUpperCase();
          }

          // Display the results in sequence text area
          sequenceArea.setText(seqText);
        }
      });
    }

    public static int getSequenceType(String sequence) throws
RESyntaxException {
        RE re = new RE("[actgnACGTN]+");
        String[] strings = re.split(sequence);
        int numbOfLettersOtherThanATGCNs = 0;

        for (int i = 0; i < strings.length; i++) {
          numbOfLettersOtherThanATGCNs                    +=
strings[i].length();
        }
        int length = sequence.length();
        int        numbOfACGTNs        =        length        -
numbOfLettersOtherThanATGCNs;

        re = new RE("[uU]+");
        strings = re.split(sequence);
        int numbOfLettersOtherThanUs = 0;

        for (int i = 0; i < strings.length; i++) {
          numbOfLettersOtherThanUs += strings[i].length();
        }
        int      numbOfUs      =      sequence.length()      -
numbOfLettersOtherThanUs;

        if (numbOfACGTNs / (double) length > SEQ_THRESHOLD) {
          return TYPE_DNA;
        } else if ((numbOfACGTNs + numbOfUs) / (double)
length > SEQ_THRESHOLD) {
          return TYPE_RNA;
        } else {
```

```
            return TYPE_PROTEIN;
        }
    }

    public static void main(String[] args) {
        SwingUtilities.invokeLater(new Runnable() {
            public void run() {
                final SwingBlast1_2 view = new SwingBlast1_2();
            }
        });
    }
}
```

Note how we have handled the creation of the GUI elements in SwingBlast version 1.2 (**Listing 2.5**):

```
public SwingBlast1_2() {
    super();
    seqFormInit();
}
```

We first created a method called `seqFormInit()` containing all the code to layout the components and then called the method in the code shown above. Earlier, for SwingBlast Version 1.1, we had instead bundled all the code within the main class (**Listing 2.1**):

```
public SwingBlast1_1() {

    setTitle(APP_NAME + " " + APP_VERSION);

    setDefaultCloseOperation(JFrame.EXIT_ON_CLOSE);

    ...
}
```

Using a separate method to build the GUI makes the code easier to read by separating the widget part from the implementation aspect.

Displaying Valid BLAST Options

The next step, now that we have accurately determined the type of sequence the user has entered in the text area, is determine which BLAST options to display for the particular type of input sequence. The purpose of this is to enable the application to automatically present only the valid BLAST algorithms appropriate for the input sequence provided by the user. Currently, if a user selects Nucleotide-nucleotide BLAST (BLASTN) on the NCBI BLAST server and supplies a protein sequence or a GenBank Id corresponding to a protein sequence, an error message pointing the mismatch is displayed; however, the BLAST server does not automatically present the valid options based on user input. Recall from **Table 2.1** that the valid BLAST options for nucleotide sequences are BLASTN, BLASTX and TBLASTX and the valid options for amino acid sequences are BLASTP and TBLASTN.

We will begin by adding the needed GUI elements to the `SwingBlast` application. The GUI elements we will need are five checkboxes for the five BLAST algorithms (BLASTN, BLASTP, BLASTX, TBLASTN and TBLASTX), a drop-down menu to select the databases to search the input sequence against and the E-value to specify the stringency of search. The application at this stage should appear as shown in **Fig. 2.18**. We will program these GUI elements to be inactivated upon launch of the application since no sequence is available for analysis. We will call this version 1.3 of the SwingBlast application.

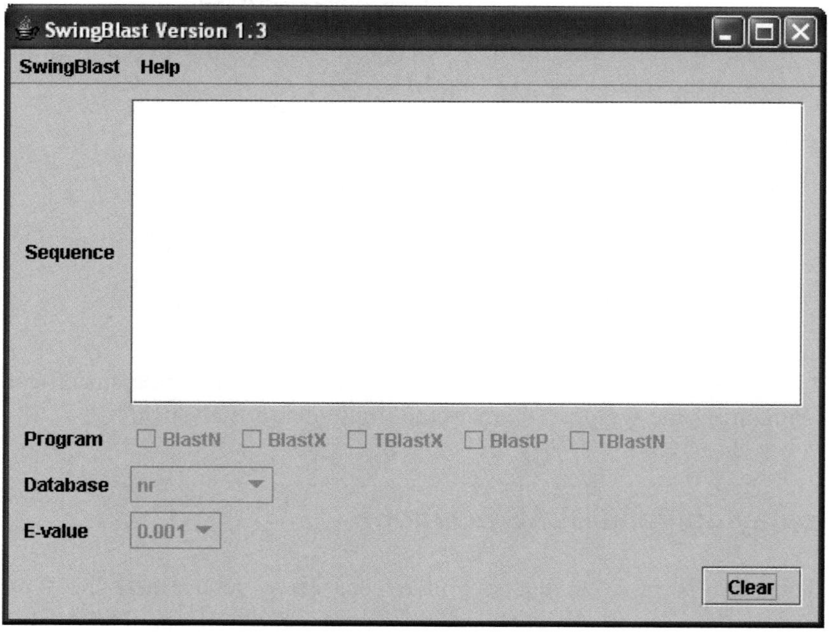

Fig. 2.18. Adding BLAST options to SwingBlast

The code to add the BLAST programs as check boxes is as follows. We first create the required array variables: BLAST_PROGRAMS_DNA, BLAST_PROGRAMS_PROTEIN, DATABASES and EVALUES to hold the appropriate allowed values for each of the parameters. Note that we are illustrating this application with a few BLAST parameters. The user can add more parameters as per individual requirements.

```
    private static final String[] BLAST_PROGRAMS_DNA = new
String[]{"BlastN", "BlastX", "TBlastX"};
    private static final String[] BLAST_PROGRAMS_PROTEIN =
new String[]{"BlastP", "TBlastN"};
    private static final String[] DATABASES = new
String[]{"nr", "est_human"};
    private static final String[] EVALUES = new
String[]{"0.001", "0.01", "0.1", "1", "10", "100"};
```

We then create the necessary widgets: check boxes for the DNA and protein BLAST options and combo boxes for the database and E-values.

```
    private JCheckBox[] cbDna;
    private JCheckBox[] cbProtein;
    private JComboBox comboDbs;
    private JComboBox comboEvalues;
```

We create a method called createProgramPanel() that draws the BLAST program panel, the database panel and the E-value panel (**Listing 2.6**).

Listing 2.6. Laying out the BLAST widgets

```
    private JPanel createProgramPanel() {
      // Create the program panel
      JPanel programPanel = new JPanel();
      JLabel program = new JLabel("Program");
      program.setPreferredSize(LABEL_PREFERRED_SIZE);
      cbDna = new JCheckBox[BLAST_PROGRAMS_DNA.length];
      String blastProgram;
      for (int i = 0; i < BLAST_PROGRAMS_DNA.length; i++) {
        blastProgram = BLAST_PROGRAMS_DNA[i];
        cbDna[i] = new JCheckBox(blastProgram);
        cbDna[i].setMaximumSize(COMBO_PREFERRED_SIZE);
      }
      cbProtein = new
JCheckBox[BLAST_PROGRAMS_PROTEIN.length];
      for (int i = 0; i < BLAST_PROGRAMS_PROTEIN.length; i++)
{
        blastProgram = BLAST_PROGRAMS_PROTEIN[i];
```

```
        cbProtein[i] = new JCheckBox(blastProgram);
        cbProtein[i].setMaximumSize(COMBO_PREFERRED_SIZE);
      }

      programPanel.setLayout(new BoxLayout(programPanel,
BoxLayout.LINE_AXIS));
      programPanel.add(program);
      programPanel.add(Box.createRigidArea(new Dimension(10,
0)));
      for (int i = 0; i < cbDna.length; i++) {
        programPanel.add(cbDna[i]);
        programPanel.add(Box.createRigidArea(new Dimension(5,
0)));
      }
      for (int i = 0; i < cbProtein.length; i++) {
        programPanel.add(cbProtein[i]);
        if (i + 1 < cbProtein.length)
          programPanel.add(Box.createRigidArea(new
Dimension(5, 0)));
      }
      programPanel.add(Box.createHorizontalGlue());
      JPanel paramPanel = new JPanel();
      paramPanel.setLayout(new BoxLayout(paramPanel,
BoxLayout.PAGE_AXIS));

      paramPanel.add(programPanel);
      paramPanel.add(Box.createRigidArea(new Dimension(0,
5)));

      // Create the database panel
      JPanel databasePanel = new JPanel();
      JLabel database = new JLabel("Database");
      database.setPreferredSize(LABEL_PREFERRED_SIZE);
      comboDbs = new JComboBox(DATABASES);
      comboDbs.setMaximumSize(COMBO_PREFERRED_SIZE);

      databasePanel.setLayout(new BoxLayout(databasePanel,
BoxLayout.LINE_AXIS));
      databasePanel.add(database);
      databasePanel.add(Box.createRigidArea(new Dimension(10,
0)));
      databasePanel.add(comboDbs);
      databasePanel.add(Box.createHorizontalGlue());
      paramPanel.add(databasePanel);
      paramPanel.add(Box.createRigidArea(new Dimension(0,
5)));

      // Create the E-Value panel
      JPanel evaluePanel = new JPanel();
      JLabel eValue = new JLabel("E-value");
      eValue.setPreferredSize(LABEL_PREFERRED_SIZE);
      comboEvalues = new JComboBox(EVALUES);
      comboEvalues.setMaximumSize(COMBO_PREFERRED_SIZE);
```

```
        evaluePanel.setLayout(new BoxLayout(evaluePanel,
BoxLayout.LINE_AXIS));
        evaluePanel.add(eValue);
        evaluePanel.add(Box.createRigidArea(new Dimension(10,
0)));
        evaluePanel.add(comboEvalues);
        evaluePanel.add(Box.createHorizontalGlue());
        paramPanel.add(evaluePanel);
        paramPanel.add(Box.createRigidArea(new Dimension(0,
5)));

        enableFunctions(TYPE_UNKNOWN);
        return paramPanel;
    }
```

The `enableFunctions()` method takes an `int` parameter (`typeOfSequence`) and is responsible for setting the check boxes for the BLAST programs to enable or disable them based on the type of sequence entered by the user. We will use the `setEnabled()` function to enable (or disable) a button. The `setEnabled()` method takes a parameter of type Boolean which can be set to `true` to enable the button and `false` to disable the button.

In case of a nucleotide sequence, we want the three check boxes for BLASTN, BLASTX and TBLASTX to be available. Simultaneously, we want the database and the E-value combo boxes to become enabled as soon as the user enters a sequence. This logic is implemented in the following manner:

```
    private void enableFunctions(int typeOfSequence) {
      if  (typeOfSequence  ==  TYPE_DNA  ||  typeOfSequence  ==
TYPE_RNA) {
        setChb(chbDna, true);
        setChb(chbProtein, false);
        setCob(cobDbs, true);
        setCob(cobEvalues, true);
      } else if (typeOfSequence == TYPE_PROTEIN) {
        setChb(chbProtein, true);
        setChb(chbDna, false);
        setCob(cobDbs, true);
        setCob(cobEvalues, true);
      } else {
        setChb(chbProtein, false);
        setChb(chbDna, false);
        setCob(cobDbs, false);
        setCob(cobEvalues, false);
      }
    }
```

In the code shown above, we define the `setChb()` and `setCob()` methods to change the settings of the check boxes (chbProtein for protein searches, chbDNA for nucleotide searches) and the combo boxes (cobDbs for database type and cobEvalues for E-values) respectively. These methods take the object type as the first parameter (check or combo box whose state needs to be set) and a `Boolean` parameter (true/false) as illustrated below:

```
    private static void setChb(JCheckBox[] boxes, boolean
value) {
      for (int i = 0; i < boxes.length; i++) {
        boxes[i].setEnabled(value);
        boxes[i].setSelected(false);
      }
    }
```

In the above method, we iterate over the check boxes, set them to enabled or disabled and ensure that they are not selected by default. For example, when the following method is called:

```
        setChb(cbDna, true);
```

the method changes only the DNA check boxes to true (enables them) since we have set cbDNA to hold the array of check boxes for only the two nucleotide related BLAST programs in the code:

```
    private static final String[] BLAST_PROGRAMS_DNA = new
String[]{"BlastN", "BlastX", "TBlastX"};

  cbDna = new JCheckBox[BLAST_PROGRAMS_DNA.length];
```

Similarly, the `setCob()` function sets the values for the combo boxes for the database and the E-values:

```
    private static void setCob(JComboBox component, boolean
value) {
      component.setEnabled(value);
      component.setSelectedIndex(0);
    }
```

Conversely, for a protein sequence, we want the BLASTP and TBLASTN check boxes and the database and the E-value combo boxes to become enabled and the check boxes for BLASTN, BLASTX and TBLASTX disabled. The method with this logic included is as follows:

```
    private void enableFunctions(int typeOfSequence) {
      if (typeOfSequence == TYPE_DNA || typeOfSequence ==
TYPE_RNA) {
        setChb(cbDna, true);
        setChb(cbProtein, false);
        setCob(comboDbs, true);
        setCob(comboEvalues, true);
      } else if (typeOfSequence == TYPE_PROTEIN) {
        setChb(cbProtein, true);
        setChb(cbDna, false);
        setCob(comboDbs, true);
        setCob(comboEvalues, true);
      } else {
        setChb(cbProtein, false);
        setChb(cbDna, false);
        setCob(comboDbs, false);
        setCob(comboEvalues, false);
      }
    }
```

We will also add a `Help` menu item. The code to add that is fairly simple:

```
      JMenu helpMenu = new JMenu("Help");
      aboutItem = new JMenuItem("About");
      helpMenu.add(aboutItem);
      menu.add(helpMenu);
```

The `Help` → `About` simply describes the current `SwingBlast` version (**Fig. 2.19**). The complete code for the application is described in **Listing 2.7**.

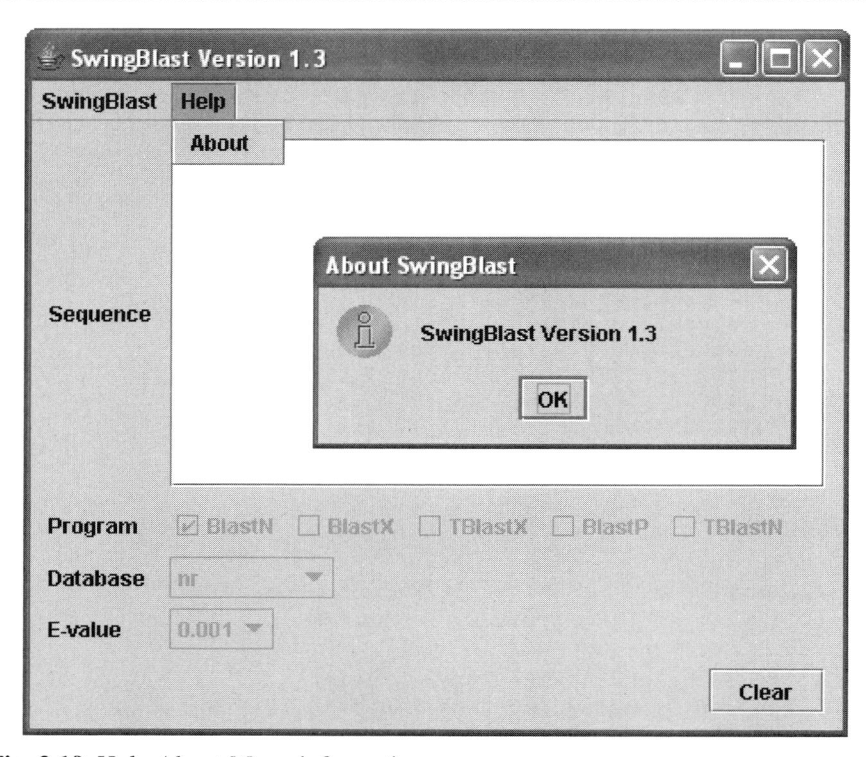

Fig. 2.19. Help About Menu information

Listing 2.7. SwingBlast version 1.3

```
package org.jfb.SwingBlast;

import org.apache.regexp.RE;
import org.apache.regexp.RESyntaxException;

import javax.swing.*;
import java.awt.*;
import java.awt.event.ActionEvent;
import java.awt.event.ActionListener;
import java.awt.event.FocusEvent;
import java.awt.event.FocusListener;

public class SwingBlast1_3 extends JFrame {
  private static final String APP_NAME = "SwingBlast";
  private static final String APP_VERSION = "Version 1.3";

  private static final Dimension LABEL_PREFERRED_SIZE = new
Dimension(57, 16);
```

```
    private static final Dimension COMBO_PREFERRED_SIZE = new
Dimension(60, 25);
    private static final Dimension CP_PREF_SIZE = new
Dimension(450, 350);

    private static final int TYPE_DNA = 0;
    private static final int TYPE_RNA = 1;
    private static final int TYPE_PROTEIN = 2;

    private static final String[] BLAST_PROGRAMS_DNA = new
String[]{"BlastN", "BlastX", "TBlastX"};
    private static final String[] BLAST_PROGRAMS_PROTEIN =
new String[]{"BlastP", "TBlastN"};
    private static final String[] DATABASES = new
String[]{"nr", "est_human"};
    private static final String[] EVALUES = new
String[]{"0.001", "0.01", "0.1", "1", "10", "100"};

    private JComponent newContentPane;
    private JTextArea sequenceArea;
    private JScrollPane scrollPaneArea;

    private JCheckBox[] chbDna;
    private JCheckBox[] chbProtein;
    private JComboBox cobDbs;
    private JComboBox cobEvalues;

    private JButton clear;

    private JMenuItem aboutItem;
    private JMenuItem quitItem;
    private static final double SEQ_THRESHOLD = 0.85;
    private static final int TYPE_UNKNOWN = -1;

    public SwingBlast1_3() {
      super();
      seqFormInit();
    }

    private void seqFormInit() {
      setTitle(APP_NAME + " " + APP_VERSION);
      setDefaultCloseOperation(JFrame.EXIT_ON_CLOSE);
      newContentPane = new JPanel();
      newContentPane.setOpaque(true);
      newContentPane.setLayout(new BorderLayout());

      setContentPane(newContentPane);

      // Create the menu bar
      JMenuBar menu = new JMenuBar();
      JMenu swingBlastMenu = new JMenu(APP_NAME);
      quitItem = new JMenuItem("Quit");
      swingBlastMenu.add(quitItem);
```

```
        menu.add(swingBlastMenu);

        JMenu helpMenu = new JMenu("Help");
        aboutItem = new JMenuItem("About");
        helpMenu.add(aboutItem);
        menu.add(helpMenu);
        setJMenuBar(menu);

        // Create the sequence pane
        JPanel sequencePanel = new JPanel();
        JLabel sequence = new JLabel("Sequence");
        sequenceArea = new JTextArea();
        sequenceArea.setLineWrap(true);
        scrollPaneArea = new JScrollPane(sequenceArea);
        scrollPaneArea.setPreferredSize(new Dimension(300,
200));

        sequencePanel.setLayout(new BoxLayout(sequencePanel,
BoxLayout.LINE_AXIS));
        sequencePanel.add(sequence);
        sequencePanel.add(Box.createRigidArea(new Dimension(10,
0)));
        sequencePanel.add(scrollPaneArea);

sequencePanel.setBorder(BorderFactory.createEmptyBorder(10,
0, 10, 0));

        // Lay out the buttons from left to right
        JPanel buttonPane = new JPanel();
        clear = new JButton("Clear");

        buttonPane.setLayout(new BoxLayout(buttonPane,
BoxLayout.LINE_AXIS));
        buttonPane.add(Box.createHorizontalGlue());
        buttonPane.add(Box.createRigidArea(new Dimension(10,
0)));
        buttonPane.add(clear);

        JPanel jPanel = new JPanel();
        jPanel.setLayout(new BorderLayout());
        jPanel.setBorder(BorderFactory.createEmptyBorder(0, 10,
10, 10));
        jPanel.add(sequencePanel, BorderLayout.NORTH);
        jPanel.add(createProgramPanel(), BorderLayout.CENTER);
        jPanel.add(buttonPane, BorderLayout.SOUTH);

        newContentPane.add(jPanel, BorderLayout.CENTER);
        newContentPane.setPreferredSize(CP_PREF_SIZE);

        // Display the window
        pack();
        Dimension screenSize =
Toolkit.getDefaultToolkit().getScreenSize();
```

```
    setLocation((screenSize.width - CP_PREF_SIZE.width) /
2,
      (screenSize.height - CP_PREF_SIZE.height) / 2);
    setVisible(true);
    addListeners();
  }

  private JPanel createProgramPanel() {
    // Create the program panel
    JPanel programPanel = new JPanel();
    JLabel program = new JLabel("Program");
    program.setPreferredSize(LABEL_PREFERRED_SIZE);
    chbDna = new JCheckBox[BLAST_PROGRAMS_DNA.length];
    String blastProgram;
    for (int i = 0; i < BLAST_PROGRAMS_DNA.length; i++) {
      blastProgram = BLAST_PROGRAMS_DNA[i];
      chbDna[i] = new JCheckBox(blastProgram);
      chbDna[i].setMaximumSize(COMBO_PREFERRED_SIZE);
    }
    chbProtein = new
JCheckBox[BLAST_PROGRAMS_PROTEIN.length];
    for (int i = 0; i < BLAST_PROGRAMS_PROTEIN.length; i++)
{
      blastProgram = BLAST_PROGRAMS_PROTEIN[i];
      chbProtein[i] = new JCheckBox(blastProgram);
      chbProtein[i].setMaximumSize(COMBO_PREFERRED_SIZE);
    }

    programPanel.setLayout(new BoxLayout(programPanel,
BoxLayout.LINE_AXIS));
    programPanel.add(program);
    programPanel.add(Box.createRigidArea(new Dimension(10,
0)));
    for (int i = 0; i < chbDna.length; i++) {
      programPanel.add(chbDna[i]);
      programPanel.add(Box.createRigidArea(new Dimension(5,
0)));
    }
    for (int i = 0; i < chbProtein.length; i++) {
      programPanel.add(chbProtein[i]);
      if (i + 1 < chbProtein.length)
        programPanel.add(Box.createRigidArea(new
Dimension(5, 0)));
    }
    programPanel.add(Box.createHorizontalGlue());
    JPanel paramPanel = new JPanel();
    paramPanel.setLayout(new BoxLayout(paramPanel,
BoxLayout.PAGE_AXIS));

    paramPanel.add(programPanel);
    paramPanel.add(Box.createRigidArea(new Dimension(0,
5)));
```

```
        // Create the database panel
        JPanel databasePanel = new JPanel();
        JLabel database = new JLabel("Database");
        database.setPreferredSize(LABEL_PREFERRED_SIZE);
        cobDbs = new JComboBox(DATABASES);
        cobDbs.setMaximumSize(COMBO_PREFERRED_SIZE);

        databasePanel.setLayout(new BoxLayout(databasePanel,
BoxLayout.LINE_AXIS));
        databasePanel.add(database);
        databasePanel.add(Box.createRigidArea(new Dimension(10,
0)));
        databasePanel.add(cobDbs);
        databasePanel.add(Box.createHorizontalGlue());
        paramPanel.add(databasePanel);
        paramPanel.add(Box.createRigidArea(new Dimension(0,
5)));

        // Create the E-Value panel
        JPanel evaluePanel = new JPanel();
        JLabel eValue = new JLabel("E-value");
        eValue.setPreferredSize(LABEL_PREFERRED_SIZE);
        cobEvalues = new JComboBox(EVALUES);
        cobEvalues.setMaximumSize(COMBO_PREFERRED_SIZE);

        evaluePanel.setLayout(new BoxLayout(evaluePanel,
BoxLayout.LINE_AXIS));
        evaluePanel.add(eValue);
        evaluePanel.add(Box.createRigidArea(new Dimension(10,
0)));
        evaluePanel.add(cobEvalues);
        evaluePanel.add(Box.createHorizontalGlue());
        paramPanel.add(evaluePanel);
        paramPanel.add(Box.createRigidArea(new Dimension(0,
5)));

        // Set it up disabled
        enableFunctions(TYPE_UNKNOWN);
        return paramPanel;
      }

      private void addListeners() {
        quitItem.addActionListener(new ActionListener() {
          public void actionPerformed(ActionEvent e) {
            System.exit(0);
          }
        });

        aboutItem.addActionListener(new ActionListener() {
          public void actionPerformed(ActionEvent e) {
            JOptionPane.showMessageDialog(SwingBlast1_3.this,
APP_NAME + " " + APP_VERSION,
                "About " + APP_NAME,
```

```
JOptionPane.INFORMATION_MESSAGE);
          }
        });

        clear.addActionListener(new ActionListener() {
          public void actionPerformed(ActionEvent e) {
            sequenceArea.setText("");
            enableFunctions(-1);
          }
        });

        sequenceArea.addFocusListener(new FocusListener() {
          public void focusGained(FocusEvent e) {
          }

          public void focusLost(FocusEvent e) {
            // Check if sequence is DNA, RNA or protein
            String text = sequenceArea.getText();

            // Format sequence in FASTA format and retrieve the
// entered sequence
            int idx = text.indexOf(">");
            boolean fastaFormatted = idx != -1;
            String seqText = null;
            String header = null;
            int seqLength = 0;
            String sequence = "";

            if (fastaFormatted) {
              int returnIdx = text.indexOf("\n");

              if (returnIdx != -1) {
                header = text.substring(0, returnIdx);
                sequence = text.substring(returnIdx + 1,
text.length()).replaceAll("\\s", "").toLowerCase();
                seqText = text;
              }
            } else {
              text = text.replaceAll("\\s", "");
              RE re = null;
              try {
                re = new RE("[0-9]+");
              } catch (RESyntaxException e1) {
                e1.printStackTrace();
              }

              boolean isGenBankID = re.match(text);

              if (isGenBankID) {
                GenbankSequenceDB genbankSequenceDB = new
GenbankSequenceDB();
                header = "GI:" + text;
                Sequence seqObject = null;
```

```
                try {
                  seqObject =
genbankSequenceDB.getSequence(text);
                  SeqIOTools.writeGenbank(System.out,
seqObject);
                } catch (Exception e1) {
                  e1.printStackTrace();
                }
                sequence = seqObject.seqString();
              } else {
                sequence = text.toLowerCase();
                header = ">Sequence1|";
                seqLength = text.length();
              }
            }

            // Check if sequence has been entered
            if (sequence.length() == 0)
              return;

            // Determine sequence type
            int typeOfSequence = TYPE_UNKNOWN;
            try {
              typeOfSequence = getSequenceType(sequence);
            } catch (RESyntaxException e1) {
              e1.printStackTrace();
            }

            String type = null;
            String unitOfLength = null;

            switch (typeOfSequence) {
              case TYPE_DNA:
                type = "DNA";
                unitOfLength = " bp";
                break;
              case TYPE_RNA:
                type = "RNA";
                unitOfLength = " bp";
                break;
              case TYPE_PROTEIN:
                type = "Protein";
                unitOfLength = " aa";
                break;
              default:
                type = "N/A";
                unitOfLength = " N/A";
            }

            if (!fastaFormatted) {
              seqText = header + type + "|" + seqLength +
unitOfLength + "\n" + sequence.toUpperCase();
```

```
            }

            // Display results
            sequenceArea.setText(seqText);

            enableFunctions(typeOfSequence);
          }
        });
      }

      private void enableFunctions(int typeOfSequence) {
        if (typeOfSequence == TYPE_DNA || typeOfSequence ==
TYPE_RNA) {
          setChb(chbDna, true);
          setChb(chbProtein, false);
          setCob(cobDbs, true);
          setCob(cobEvalues, true);
        } else if (typeOfSequence == TYPE_PROTEIN) {
          setChb(chbProtein, true);
          setChb(chbDna, false);
          setCob(cobDbs, true);
          setCob(cobEvalues, true);
        } else {
          setChb(chbProtein, false);
          setChb(chbDna, false);
          setCob(cobDbs, false);
          setCob(cobEvalues, false);
        }
      }

      private static void setChb(JCheckBox[] boxes, boolean
value) {
        for (int i = 0; i < boxes.length; i++) {
          boxes[i].setEnabled(value);
        }
      }

      private static void setCob(JComponent component, boolean
value) {
        component.setEnabled(value);
      }

      public static int getSequenceType(String sequence) throws
RESyntaxException {
        RE re = new RE("[actgnACGTN]+");
        String[] strings = re.split(sequence);
        int numbOfLettersOtherThanATGCNs = 0;

        for (int i = 0; i < strings.length; i++) {
          numbOfLettersOtherThanATGCNs += strings[i].length();
        }
        int length = sequence.length();
        int numbOfACGTNs = length -
```

```
numbOfLettersOtherThanATGCNs;

        re = new RE("[uU]+");
        strings = re.split(sequence);
        int numbOfLettersOtherThanUs = 0;

        for (int i = 0; i < strings.length; i++) {
          numbOfLettersOtherThanUs += strings[i].length();
        }
        int numbOfUs = sequence.length() -
numbOfLettersOtherThanUs;

        if (numbOfACGTNs / (double) length > SEQ_THRESHOLD) {
          return TYPE_DNA;
        } else if ((numbOfACGTNs + numbOfUs) / (double) length
> SEQ_THRESHOLD) {
          return TYPE_RNA;
        } else {
          return TYPE_PROTEIN;
        }
      }

      public static void main(String[] args) {
        SwingUtilities.invokeLater(new Runnable() {
          public void run() {
            final SwingBlast1_3 view = new SwingBlast1_3();
          }
        });
      }
    }
```

Fig. 2.20 and **Fig. 2.21** show the behavior of the application for a nucleotide and a protein sequence respectively that is entered in the text area. In both cases, the correct set of BLAST programs are selected (BLASTN, BLASTX and TBLASTX for nucleotide sequence and BLASTP and TBLASTN for protein sequence). Simultaneously, the drop-down menu boxes for the databases and the E-value are activated for selection by the user.

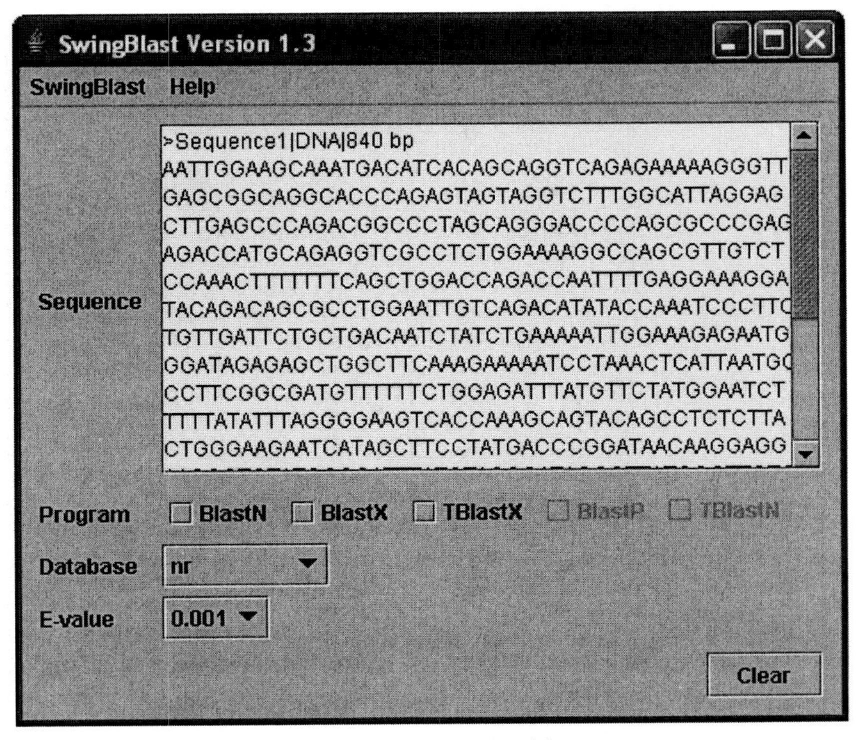

Fig. 2.20. Displaying BLAST options for a nucleotide sequence

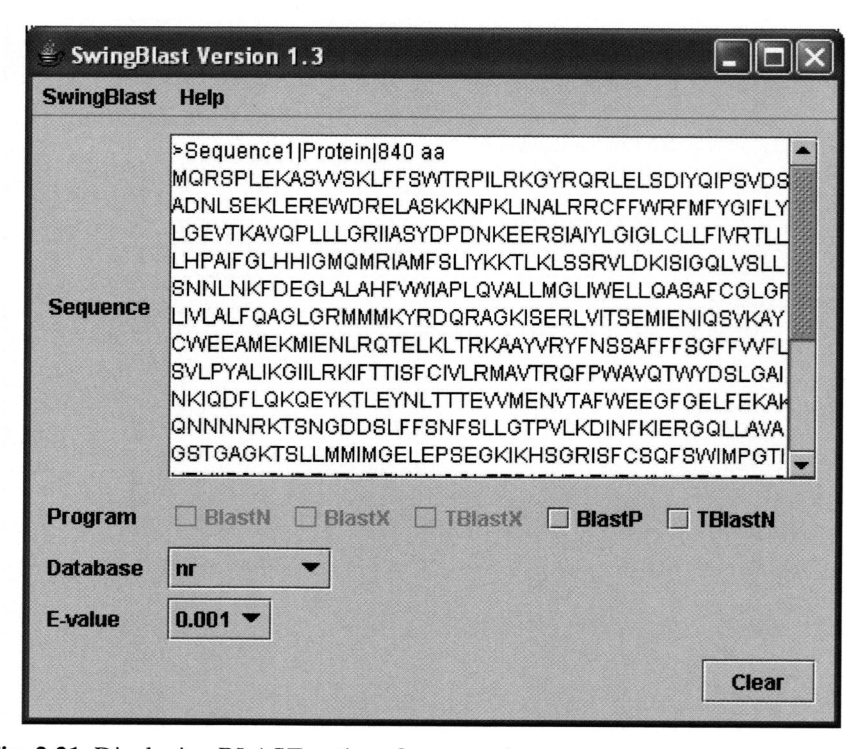

Fig. 2.21. Displaying BLAST options for a protein sequence

Summary

In this Chapter, we created a Swing based application that allows users to prepare sequences for BLAST searches by performing simple formatting tasks such as conversion into the Fasta format and determining the sequence type and length. Along the way we introduced how to write code to respond to events taking place in response to user initiated actions. We created the GUI elements and wrote the code that enables the elements to respond to the sequence type and present only the valid BLAST options that are available for the entered sequence type. The rationale for building these features into the application was to make it more functional and to simplify its use for the end-users, given the many potentially confusing parameters a user has to supply when performing a search operation. In the next Chapter, we will extend the `SwingBlast` application to actually perform the BLAST search operation.

Questions and Exercises

1. Enhance the SwingBlast application interface to accept multiple sequences, for example, by incorporating the ability to upload a multiple Fasta file. Next incorporate code to add checkboxes against each uploaded sequence to allow users to select specific sequences for further analysis. Develop the use cases that fulfill the above user requirements.

2. Explore the BLAST algorithms in further detail by visiting the tutorial site listed below. How do you determine the statistical significance of BLAST hits? What are bit scores and p-values?

3. Download the sequence for simian sarcoma virus v-sis oncogene gene from GenBank and perform a BLAST against the nr database. What BLAST program(s) would you use to find similarities between v-sis and existing nucleotide and protein sequences? What are the top ten hits that BLAST returns? Which human and other vertebrate homologs can you identify?

Additional Resources

- BLAST tutorial - http://www.ncbi.nlm.nih.gov/BLAST/tutorial/Altschul-1.html
- GenBank - http://www.ncbi.nlm.nih.gov/Genbank/index.html
- Java™ 2 Platform Standard Edition 5.0 API Specification - http://java.sun.com/j2se/1.5.0/docs/api/

Selected Reading

Simian sarcoma virus onc gene, v-sis, is derived from the gene (or genes) encoding a platelet-derived growth factor. Doolittle RF, Hunkapiller MW, Hood LE, Devare SG, Robbins KC, Aaronson SA, Antoniades HN. Science. 1983 Jul 15;221(4607):275–277.

Identification of the cystic fibrosis gene: cloning and characterization of complementary DNA. Riordan JR, Rommens JM, Kerem B, Alon N,

Rozmahel R, Grzelczak Z, Zielenski J, Lok S, Plavsic N, Chou JL, et al. Science. 1989 Sep 8;245(4922):1066-73.

Basic local alignment search tool. Altschul SF, Gish W, Miller W, Myers EW, Lipman DJ. J Mol Biol. 1990 Oct 5;215(3):403-10.

Gapped BLAST and PSI-BLAST: a new generation of protein database search programs. Altschul SF, Madden TL, Schaffer AA, Zhang J, Zhang Z, Miller W, Lipman DJ. Nucleic Acids Res. 1997 Sep 1;25(17):3389-402.

Chapter III

Running BLAST using SwingBlast

Introduction

In the last Chapter, we created the basic framework application called `SwingBlast` Version 1.3 using Swing libraries to manipulate user defined nucleotide and protein sequences and prepare them for BLAST searches. In this Chapter, we will add functionality to the application that enables users to download sequences automatically from NCBI GenBank, submit sequences for multiple simultaneous BLAST analyses, and save and view BLAST results. To begin with, we will demonstrate how to use NCBI's *QBlast* package to perform BLAST searches. We will then create an application called JQBlast to demonstrate how to use theQBlast package to run BLAST searches.

The NCBI QBLAST Package

NCBI provides a standardized API called URLAPI to formulate and dispatch direct HTTP-encoded requests to the NCBI QBlast system. The URLAPI provides a URL and a mechanism to set parameters that allows users to send sequences for BLAST searches.

NCBI QBlast works through 4 steps:

1. The user provides BLAST parameters through a URL using the HTTP POST method

2. The QBlast service returns a Request Identifier (RID) and a Request Time of Execution (RTOE, measured in seconds) for the search, which provide respectively, a unique identifier for the search operation and an estimate of the time required to complete the search

3. The user queries the QBlast service with the RID through HTTP GET method

4. The server sends back the result with a status value that indicates the progress of the BLAST request

Users of the QBlast service should adhere to the guidelines provided by NCBI when submitting large batch searches. In general, searches should be performed in a sequential manner after receiving the RID and RTOE for each submission. NCBI specifies that each request be submitted after a pause of no less than 3 seconds to check on the status of the request using the RID. Failure to do so may overload the server and force NCBI to block offending users from further use of the service.

Strategy for Creating a QBlast Based System

The design of the NCBI QBlast service as described above stipulates the need for a client application that performs the following operations:

1. Send search requests made by the user and check the status of requests periodically

2. Perform the appropriate action based on the nature of the status value that gets returned

QBlast may return one of three types of status values: "READY" meaing that the search was completed successfully, "WAITING" meaning that the search has not been completed and "UNKNOWN" meaning that an error has been encountered during the BLAST submission and/or search process. In UML terms, the user and the client application are actors that interact with the QBlast system. The UML diagram below (**Fig. 3.1**) depicts the use cases that encapsulate the basic functionality that is desired of the system that we wish to create:

1. User submits query sequence to the QBlast service

2. Application queries status of the BLAST search with a unique RID

3. Applcation returns results approriate to the status value

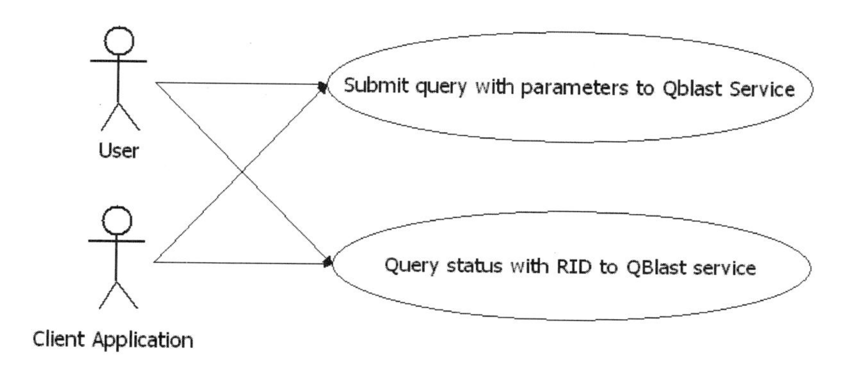

Fig. 3.1. Use Cases for the QBlast service

In terms of the architecture of the application, we will provide a class that will wrap the NCBI URLAPI into Java API that can be reused in other applications. To fulfill these use cases, we will design the QBlast service to implement 2 methods: `submitQuery` and `queryStatus` (**Fig. 3.2**).

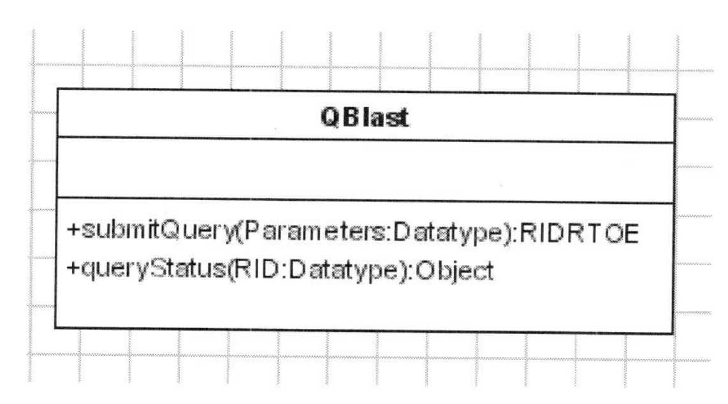

Fig. 3.2. Class QBlast

The `QBlast` class is our interface to the real NCBI URLAPI. From the application point of view, it is totally transparent and is designed to be so in order to accommodate and simplify future changes to the API (or, if there is a need to adopt an entirely different API). This design ensures that the framework we create remains usable even if the underlying API requires changes. The `submitQuery()` method takes the BLAST parameters (specified through the `QBlastParameter` object) and returns an object of type `RequestIdentifier`. The parameters needed to run the BLAST search would be obtained from the user through the `SwingBlast` GUI we created in Chapter 1. The `RequestIdentifier` is returned by the QBlast service in response to the submitted request and contains the RID and the RTOE for a specific search.

For the `queryStatus` method similarly, we will need 2 objects: `RequestIdentifier` and `QBlastResult`. A UML diagram with these considerations in mind is shown in **Fig. 3.3.**

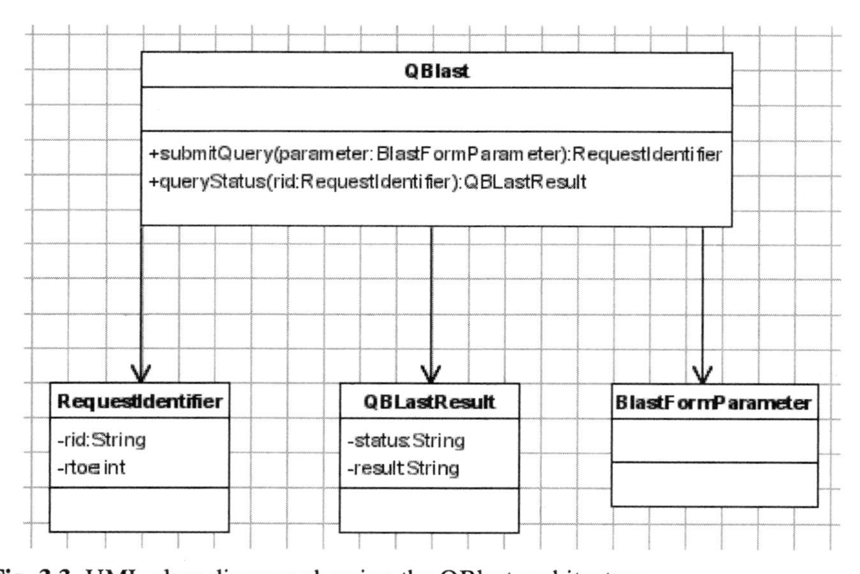

Fig. 3.3. UML class diagram showing the QBlast architecture

Designing the BLAST API

We will design our BLAST API to consist of 3 classes:

- `Blast`
- `BlastManager`
- `BlastException`

We will define `Blast` as an *abstract class*, which means that it represents an abstract concept, and therefore cannot be instantiated, but can only be subclassed. An abstract class is declared using the keyword *abstract* before the class keyword in the class declaration. In this case, for example, we would declare the *Blast* class as shown below:

```
abstract class blast { … }
```

We'll describe this class in detail later in the Chapter. The *BlastManager* class provides a mechanism to get an instance of the abstract class *Blast* without having to worry about how to create the instance by calling the *static* method (that we had earlier explained in Chapter 2):

```
Blast blast = BlastManager.createBlast();
```

The `BlastException` class provides a mechanism for handling exceptions thrown by any implementation when a failure or error occurs. The `RequestIdentifier` class is a Java class, which provides what are known as *setter*, and *getter* methods that provide information about the request submitted to the Blast service. What are *setter* and *getter* methods? In a class definition, private fields can be encapsulated so that the data structure used can be changed at will without compromising the rest of the code that uses that class. When the data structure is hidden, the way to provide access to and/or modify the fields is through setter and getter methods. For example, a class that has a field called `result` will provide a *setter* method called `setResult` and a *getter* method call `getResult`. The `RequestIdentifier` class uses these methods as described above. The structure of the application designed so far is shown in **Fig. 3.4**.

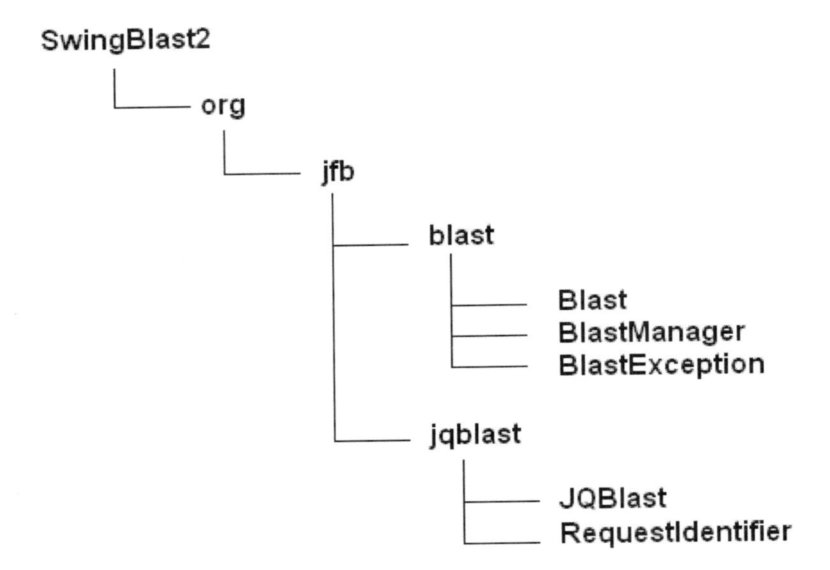

Fig. 3.4. Structure of the SwingBlast application

Description of Blast Classes

The `Blast` class extends the *Observable* class, which represents an observable object, an instance of which can be observed for any changes that occur to the object. When an observable instance changes (that is, when an object that is being observed changes), the `notifyObservers` method is called and causes the observer to be notified of the change by a call to the observer's update method. In this case, we want to observe the `Blast` class for changes that occur during the process of submitting the request and waiting for the result to be returned. We can then notify the observers of the progress of the search, as well as when the results are ready or if an error occurs.

The `Blast` class contains 2 abstract methods:

```
submitQuery()
```

and,

```
requestResult()
```

that respectively take one parameter each: a Java data type called *Map* for the BLAST parameters, which stores sets of elements in the form of key-value pairs, and the identifier for the identifier that is uniquely associated with each BLAST search. Both methods return an object of the respective type and throw an exception if an error occurs. The `Blast` class is defined in **Listing 3.1**.

Listing 3.1. The Blast class

```
package org.jfb.blast;

import java.util.HashMap;
import java.util.Observable;

public abstract class Blast extends Observable {
public abstract Object submitQuery(Map parameters)
throws BlastException;

  public abstract Object requestResult(Object identifier)
throws BlastException;
}
```

The way to initialize the `BlastManager` is to provide the full class name of the implementation through a *JVM system property* called '`blast.driver`'. An example of how to provide our BLAST implementation called `org.jfb.jqblast.JQBlast` to the BlastManager via a JVM system property is shown below:

```
Java -Dblast.driver=org.jfb.jqblast.JQBlast ...
```

The class `JQBlast` must be declared in the *Java classpath* to be able to be found by the *Java classloader*. The Java classloader is responsible for loading a Java class when it is needed. The `BlastManager` class is described in **Listing 3.2**.

Listing 3.2. The BlastManager class

```
package org.jfb.blast;

public class BlastManager {
    private static String blastClass = null;
    private static boolean initialized = false;

    public static void register(Blast blast) {
        blastClass = blast.getClass().getName();
    }
```

```
      private static void loadInitialDrivers() {
           final           String           driver             =
System.getProperty("blast.driver");
           if (driver == null)
               return;

           try {
               System.out.println("BlastManager.Initialize:
loading " + driver);
               Class.forName(driver);
           } catch (Exception e) {
               System.out.println("BlastManager.Initialize:
load failed: " + e);
           }
      }

      public static Blast createBlast() throws BlastException
{
           if (!initialized) {
               initialized = true;
               loadInitialDrivers();
           }
           if (blastClass == null)
               throw new BlastException("There is no driver
configured! "
                        +   "Please  use  blast.driver  Java
property or Class.forName to load the driver class.");
           try {
               // In a multi thread environment we need to
make sure // that the class is loaded.
               final     Class     aClass     =     (Class)
Class.forName(blastClass,                            true,
Thread.currentThread().getContextClassLoader());
               return    (Blast)   aClass.getConstructor(new
Class[]{}).newInstance(new Object[]{});
           } catch (Exception e) {
               throw new BlastException(e);
           }
      }
  }
```

The purpose of the register() method is to inform the BlastManager which Blast implementation we want to use. This is done as follows:

```
      public static void register(Blast blast) {

          blastClass = blast.getClass().getName();

      }
```

Here, `blast.getClass()` returns an instance of class `java.lang.Class`. `blastClass` is an instance of `java.lang.Class` and `blastClass.getName()` will return the real class name which, in this case would be `org.jfb.jqblast.JQBlast`.

Let's look at the `loadInitialDrivers` method below:

```
private static void loadInitialDrivers() {
        final           String          driver          =
System.getProperty("blast.driver");
        if (driver == null)
            return;

        try {
            System.out.println("BlastManager.Initialize:
loading " + driver);
            Class.forName(driver);
        } catch (Exception e) {
            System.out.println("BlastManager.Initialize:
load failed: " + e);
        }
    }
```

When the `loadInitialDrivers` method is called, it gets the property `blast.driver` from the system and if it is not null, calls the `Class.forName()` method. At that point, `BlastManager` knows that a `Blast` driver is registered and available, otherwise an exception is thrown with an error message. Finally, the `BlastException` class handles any exceptions that arise during the BLAST search (**Listing 3.3**).

Listing 3.3. The BlastException class

```
package org.jfb.blast;

public class BlastException extends Exception {
    public BlastException() {
    }

    public BlastException(String message) {
        super(message);
    }

    public BlastException(String message, Throwable cause)
{
        super(message, cause);
    }

    public BlastException(Throwable cause) {
```

```
            super(cause);
        }
    }
```

Implementing JQBlast

We will now build the `JQBlast` application that allows users to send multiple simultaneously BLAST queries using the classes we described above. To implement the NCBI QBlast package, we just need to extend the `Blast` class and provide an implementation of the methods as described above. We will call the instance of the `Blast` class `JQBlast` as shown below:

```
public class JQBlast extends Blast {
    //implement Blast methods
}
```

We will create a file called `QBlast.java` to implement this code. It is up to the developer of a `Blast` implementation to provide the code for those methods. The developer must also register the `Blast` class to the `BlastManager` class using a *static* statement that will be executed after loading the class. A *static* statement is a piece of code that starts with the Java keyword *static* and is followed by curly brackets (which, in this case, holds the code that loads the Blast implementation called `org.jfb.jqblast.JQBlast`). It is executed after the class is loaded in the JVM:

```
public class JQBlast extends Blast {
static {
        System.out.println("Registering " + JQBlast.class);
        BlastManager.register(new JQBlast());
        }
    //implement Blast methods
    }
```

The `Blast` engine provides a mechanism to specify the parameters for a search (such as database type, BLAST algorithm type, E-value, etc.) and to submit a sequence into a queue for the actual `Blast` operation. The above design provides a way of accessing an instance of `Blast`, without the need to know the mechanism by which the `Blast` operation is submitted or performed. In this case, `JQBlast` is an implementation of the abstract `Blast` class and that is the one that is instantiated by the `BlastManager`.

When a Java class is loaded, the *Java classloader* will run all the *static* statements first, so a `JQBlast` instance will be created and registered to the `BlastManager`. Now to allow the classloader to load that class we need to call the *java classpath* using the `forName` method from the class `Class`, as shown below:

```
static {
        try {
            Class.forName("org.jfb.jqblast.JQBlast");
        } catch (ClassNotFoundException e) {
            e.printStackTrace();
        }
    }
```

Alternately, we can pass the Java class name to the JVM system property using the Java *–D* option and the property name "`blast.driver`", if we don't want to hard code the name of the `Blast` class we would like to use in the code.

```
Java -Dblast.driver=org.jfb.SwingBlast.qblast.QBlast (…)
```

The property is then retrieved using the `getProperty` method as shown below:

```
System.getProperty("blast.driver");
```

We pass the BLAST parameters to the `submitQuery()` method as follows:

```
public    Object    submitQuery(Map    parameters)    throws
BlastException {
          String urlapiQuery = createUrlapiQuery(parameters);
          setChanged();
          notifyObservers("Submitting the job to the server
with query\n"
  + urlapiQuery);
          String queryResult = sendQuery(urlapiQuery);

          if (queryResult == null) return null;
          return parseOutReqId(queryResult);
    }
```

The method `createUrlapiQuery()` within `submitQuery()` generates the HTTP-encoded request containing the specified parameters (including the sequence specified by the user (in this case, a test sequence "AAGTCGATAGCTCGCGCGCCGGCCGTGAGGAAAAAAAAA").

```
CMD=Put&QUERY_BELIEVE_DEFLINE=yes&QUERY=%3E+Sequence1%7CDNA
%7C38+bp%0AAGTCGATAGCTCGCGCGCCGGCCGTGAGGAAAAAAAAA&DATABASE=nr
&PROGRAM=blastn&EXPECT=0.001
```

The method is described below:

```
    private String createUrlapiQuery(Map parameters) {
        StringBuffer          query          =          new
StringBuffer("CMD=Put&QUERY_BELIEVE_DEFLINE=yes");
        try {

query.append("&QUERY=").append(URLEncoder.encode((String)
parameters.get("sequenceText"), "UTF-8"))
                    .append("&DATABASE=").append((String)
parameters.get("database"))
                    .append("&PROGRAM=").append((String)
parameters.get("blastType"))
                    .append("&EXPECT=").append((String)
parameters.get("eValue"));
        } catch (UnsupportedEncodingException uee) {
            uee.printStackTrace();
        }
        return query.toString();
    }
```

In this case, the method returns a String object containing the sequence to be submitted for the BLAST search, the database to be searched against, the BLAST program to be used and the cut-off E-value for the search.

The `setChanged()` method in `submitQuery()` is derived from the `Observable` class and is used to keep track of changes in the status of an object, in this case, `Blast`. The `Observable` class notifies changes in states of objects by calling the `notifyObservers()` method. In this example, we will inform the user that a search job has been submitted (with the message, "Submitting the job to the server with query", and appends the `urlapiQuery` string to it.

```
    notifyObservers("Submitting the job to the server with
query\n" + urlapiQuery);
```

Next we send the query for BLAST using the `sendQuery()` method:

```
  private   String   sendQuery(String   httpQuery)   throws
BlastException {
    DataOutputStream printer = null;
    URLConnection urlConnection;
    ByteArrayOutputStream outputStream = null;
```

```
      try {
        urlConnection = new URL(blastUrl).openConnection();
        urlConnection.setDoInput(true);
        urlConnection.setDoOutput(true);
        urlConnection.setUseCaches(false);
        urlConnection.setRequestProperty("Content-Type",
"application/x-www-form-urlencoded");
        urlConnection.setRequestProperty("Content-Length", ""
+ httpQuery.length());
        printer                    =                    new
DataOutputStream(urlConnection.getOutputStream());
        printer.writeBytes(httpQuery);

        // Read the result
        BufferedReader reader = null;
        reader = new BufferedReader(new
          InputStreamReader(urlConnection.getInputStream()));
        outputStream = new ByteArrayOutputStream();
        String str;
        while ((str = reader.readLine()) != null) {
          outputStream.write(str.getBytes());
        }
      } catch (MalformedURLException mue) {
        mue.printStackTrace();
        throw new BlastException(blastUrl + " is malformed");
      } catch (IOException ioe1) {
        ioe1.printStackTrace();
        throw    new    BlastException("Could    not    get    the
connection or write to it");
      } finally {
        try {
          printer.close();
          printer = null;
        } catch (IOException ignore) {
          ignore.printStackTrace();
        }
      }
      return outputStream != null ? outputStream.toString() :
null;
    }
```

The `sendQuery()` method returns a String carrying the results of the operation:

```
  String queryResult = sendQuery(urlapiQuery);
```

We then parse the result (unless no hits were found) using the `parseOutReqId()` method:

```
      if (queryResult == null) return null;
```

```
    return parseOutReqId(queryResult);
```

The `parseOutReqId()` method parses the RID and RTOE from the returned string which is of type:

```
QBlastInfoBegin                    RID    =    1097884888-2134-
17842894979.BLASTQ4   RTOE = 30QBlastInfoEnd
```

and returns the `RequestIdentifier`:

```
private RequestIdentifier parseOutReqId(String string) {
    String rid = null;
    String rtoe = null;

    try {
      RE                regex                    =                new
RE("QBlastInfoBegin(\\s*)RID(\\s*)=\\2(\\S*)(\\s*)RTOE\\2=\\2
(.*)QBlastInfoEnd");
      boolean matched = regex.match(string);

      if (matched) {
        rid = regex.getParen(3);
        rtoe = regex.getParen(5);
      }
    } catch (RESyntaxException ree) {
    }
    if (rid == null || rtoe == null)
      return null;
    return              new              RequestIdentifier(rid,
Integer.parseInt(rtoe));
  }
```

Once we obtain the RID and RTOE, we wait for a period of time specified by the RTOE before trying to access the results.

```
public  Object  requestResult(RequestIdentifier  identifier)
throws BlastException {
    if (identifier == null)
      throw  new  BlastException("Cannot  get  the  request
identifier");

    setChanged();
    notifyObservers("Getting  from  JQBlast  Service  the  RID
("   +   identifier.getRid()   +   ")   and   RTOE   ("   +
identifier.getRtoe() + ").");

    // Wait  the  rtoe  time  before  sending  any  request  back
to the server
    try {
      long     timeOut     =     identifier.getRtoe()     +
```

```
identifier.getTime();

        if (timeOut > System.currentTimeMillis()) {
          int     timeLeft     =     ((int)     (timeOut     -
System.currentTimeMillis())) * 1000;

          synchronized (this) {
            while (timeLeft > 0) {
              wait(waitTime);
              setChanged();
              notifyObservers("Time  left  "  +  ((timeLeft  -=
waitTime) / 1000) + "s before requesting the result");
            }
          }
        }

      } catch (InterruptedException ie) {
        ie.printStackTrace();
      }

      setChanged();
      notifyObservers("Requesting  the  result  for  rid:  "  +
identifier.getRid());
      StringBuffer             query             =             new
StringBuffer("CMD=Get&FORMAT_TYPE=XML");
      query.append("&RID=" + identifier.getRid());
      String ri = query.toString();

      String queryResult = null;
      String status = null;

      boolean hasResult = false;
      int ct = 0;
      RE regex = null;
      try {
        regex                        =                        new
RE("QBlastInfoBegin(\\s*)Status=(.*)QBlastInfoEnd");
      } catch (RESyntaxException ree) {

      }

      synchronized (this) {
        while (!hasResult) {
          status = null;
          queryResult = sendQuery(ri);
          boolean matched = regex.match(queryResult);

          if (matched) {
            status = regex.getParen(2);
          }
          hasResult = !"WAITING".equals(status);
          if (hasResult) {
            break;
```

```
        }

        setChanged();
        notifyObservers("Waiting " + NUMBER_OF_SECOND + "
seconds before re-trying (total waiting time: " + (ct +=
NUMBER_OF_SECOND) + "s).");
        try {
          wait(NUMBER_OF_SECOND * 1000);
        } catch (InterruptedException ie1) {
          ie1.printStackTrace();
        }
      }
    }
    if ("UNKNOWN".equals(status)) {
      throw new BlastException("Result for RID " +
identifier.getRid() + " failed.");
    }
    setChanged();
    notifyObservers("Getting back the blast result in
XML");
    return queryResult;
  }
```

The complete code for `JQBlast.java` is shown in **Listing 3.4**.

Listing 3.4. JQBlast.java

```
package org.jfb.jqblast;

import org.apache.regexp.RE;
import org.apache.regexp.RESyntaxException;
import org.jfb.blast.Blast;
import org.jfb.blast.BlastException;
import org.jfb.blast.BlastManager;

import java.io.BufferedReader;
import java.io.ByteArrayOutputStream;
import java.io.DataOutputStream;
import java.io.File;
import java.io.FileOutputStream;
import java.io.IOException;
import java.io.InputStreamReader;
import java.io.OutputStream;
import java.io.UnsupportedEncodingException;
import java.net.MalformedURLException;
import java.net.URL;
import java.net.URLConnection;
import java.net.URLEncoder;
import java.util.HashMap;

public class JQBlast extends Blast {
```

```
      static {
          System.out.println("Registering " + JQBlast.class);
          BlastManager.register(new JQBlast());
      }

      private    static    final    String    blastUrl    =
"http://www.ncbi.nlm.nih.gov/blast/Blast.cgi";
      private static final int NUMBER_OF_SECOND = 3;

      public   Object   submitQuery(Map   parameters)   throws
BlastException {
          String urlapiQuery = createUrlapiQuery(parameters);
          setChanged();
          notifyObservers("Submitting the job to the server
with query\n" + urlapiQuery);
          String queryResult = sendQuery(urlapiQuery);

          if (queryResult == null) return null;
          return parseOutReqId(queryResult);
      }

      final static int waitTime = 2000;

      public  Object  requestResult(Object  identifier)  throws
BlastException {
          if  (identifier  ==  null  ||  !(identifier instanceof
RequestIdentifier))
              throw   new   BlastException("Cannot   get   the
request identifier " + identifier);

          RequestIdentifier rIdentifier = (RequestIdentifier)
identifier;
          setChanged();
          notifyObservers("Getting from JQBlast Service the
RID (" + rIdentifier.getRid()
                  + ") and RTOE (" + rIdentifier.getRtoe() +
").");

          // Wait the rtoe time before sending any request
back to the server
          try {
              long    timeOut    =    rIdentifier.getRtoe()    +
rIdentifier.getTime();

              if (timeOut > System.currentTimeMillis()) {
                  int    timeLeft    =    ((int)    (timeOut    -
System.currentTimeMillis())) * 1000;

                  synchronized (this) {
                      while (timeLeft > 0) {
                          wait(waitTime);
                          setChanged();
                          notifyObservers("Time   left   "   +
```

```
((timeLeft -= waitTime) / 1000) + "s before requesting the
result");
                    }
                }
            }

        } catch (InterruptedException ie) {
            ie.printStackTrace();
        }

        // do a loop every 3 seconds send the request until
we get the status = READY and the blast result
        // End of loop
        setChanged();
        notifyObservers("Requesting the result for rid: " +
rIdentifier.getRid());
        StringBuffer          query          =          new
StringBuffer("CMD=Get&FORMAT_TYPE=XML");
        query.append("&RID=" + rIdentifier.getRid());
        String ri = query.toString();

        String queryResult = null;
        String status = null;

        boolean hasResult = false;
        int ct = 0;
        RE regex = null;
        try {
            regex              =              new
RE("QBlastInfoBegin(\\s*)Status=(.*)QBlastInfoEnd");
        } catch (RESyntaxException ree) {
            // We ignore it since we've checked the regex
already!
        }
        Runtime runtime = Runtime.getRuntime();

        synchronized (this) {
            while (!hasResult) {
                status = null;
                queryResult = sendQuery(ri);
                boolean matched = regex.match(queryResult);

                if (matched) {
                    status = regex.getParen(2);
                }
                hasResult = !"WAITING".equals(status);
                if (hasResult) {
                    break;
                }

                setChanged();
                notifyObservers("Waiting          "          +
NUMBER_OF_SECOND
```

```
                          + " seconds before re-trying (total
waiting time: " + (ct += NUMBER_OF_SECOND) + "s). "
                          + runtime.freeMemory() + " bytes
left");
                    try {
                        wait(NUMBER_OF_SECOND * 1000);
                    } catch (InterruptedException ie1) {
                        ie1.printStackTrace();
                    }
                }
            }
            if ("UNKNOWN".equals(status)) {
                throw new BlastException("Result for RID " +
rIdentifier.getRid() + " failed.");
            }
            setChanged();
            String fileName = createTempFileName();
            try {
                OutputStream outputStream = new
FileOutputStream(fileName);
                outputStream.write(queryResult.getBytes());
            } catch (IOException ioe) {
                throw new BlastException("Saving result for RID
" + rIdentifier.getRid()
                        + " into " + fileName + " failed.",
ioe);
            }
            notifyObservers("Getting back the blast result in
XML " + queryResult.length());
            return fileName;
        }

        private String sendQuery(String httpQuery) throws
BlastException {
            DataOutputStream printer = null;
            URLConnection urlConnection;
            ByteArrayOutputStream outputStream = null;
            String fileName = null;

            try {
                urlConnection = new
URL(blastUrl).openConnection();
                urlConnection.setDoInput(true);
                urlConnection.setDoOutput(true);
                urlConnection.setUseCaches(false);
                urlConnection.setRequestProperty("Content-
Type", "application/x-www-form-urlencoded");
                urlConnection.setRequestProperty("Content-
Length", "" + httpQuery.length());
                printer = new
DataOutputStream(urlConnection.getOutputStream());
                printer.writeBytes(httpQuery);
```

```
                // Let's read the result
                BufferedReader reader = null;
                reader        =        new        BufferedReader(new
InputStreamReader(urlConnection.getInputStream()));
                outputStream = new ByteArrayOutputStream();
                String str;
                while ((str = reader.readLine()) != null) {
                    outputStream.write(str.getBytes());
                }
            } catch (MalformedURLException mue) {
                mue.printStackTrace();
                throw   new   BlastException(blastUrl   +   "   is
malformed");
            } catch (IOException ioe1) {
                ioe1.printStackTrace();
                throw  new  BlastException("Could  not  get  the
connection or write to it");
            } finally {
                try {
                    printer.close();
                    printer = null;
                } catch (IOException ignore) {
                    ignore.printStackTrace();
                }
            }
            return   outputStream   ==   null   ?   null   :
outputStream.toString();
        }

        private RequestIdentifier parseOutReqId(String string)
{
            String rid = null;
            String rtoe = null;

            try {
                // <!--QBlastInfoBegin      RID  =  1097884888-
2134-17842894979.BLASTQ4    RTOE = 30QBlastInfoEnd-->
                RE            regex            =            new
RE("QBlastInfoBegin(\\s*)RID(\\s*)=\\2(\\S*)(\\s*)RTOE\\2=\\2
(.*)QBlastInfoEnd");
                boolean matched = regex.match(string);

                if (matched) {
                    rid = regex.getParen(3);
                    rtoe = regex.getParen(5);
                }
            } catch (RESyntaxException ree) {
                // We  ignore  it  since  we  checked  the  regex
already!
            }
            if (rid == null || rtoe == null)
                return null;
            return          new          RequestIdentifier(rid,
```

```
Integer.parseInt(rtoe));
        }

        private String createUrlapiQuery(Map parameters) {
            StringBuffer            query           =           new
StringBuffer("CMD=Put&QUERY_BELIEVE_DEFLINE=yes");
            try {

query.append("&QUERY=").append(URLEncoder.encode((String)
parameters.get("sequenceText"), "UTF-8"))
                        .append("&DATABASE=").append((String)
parameters.get("database"))
                        .append("&PROGRAM=").append((String)
parameters.get("blastType"))
                        .append("&EXPECT=").append((String)
parameters.get("eValue"));
            } catch (UnsupportedEncodingException uee) {
                uee.printStackTrace();
            }
            return query.toString();
        }

        private String createTempFileName() {
            return    System.getProperty("java.io.tmpdir")    +
File.separator
                    + "blast-" + System.currentTimeMillis() +
".xml";

        }
        private static String packBy(int i, String s) throws
RESyntaxException {
            String substIn = "[a-zA-Z]{" + i + "}";
            String substTo = "$0 ";
            RE re = new RE(substIn);
            return              re.subst(s,              substTo,
RE.REPLACE_BACKREFERENCES);
        }
    }
```

Enhancing the SwingBlast Application

Let's also take a look at the code that generates the GUI for the application. The `SwingBlast` Version 1.3 we created in the last Chapter is shown in **Fig. 3.5**.

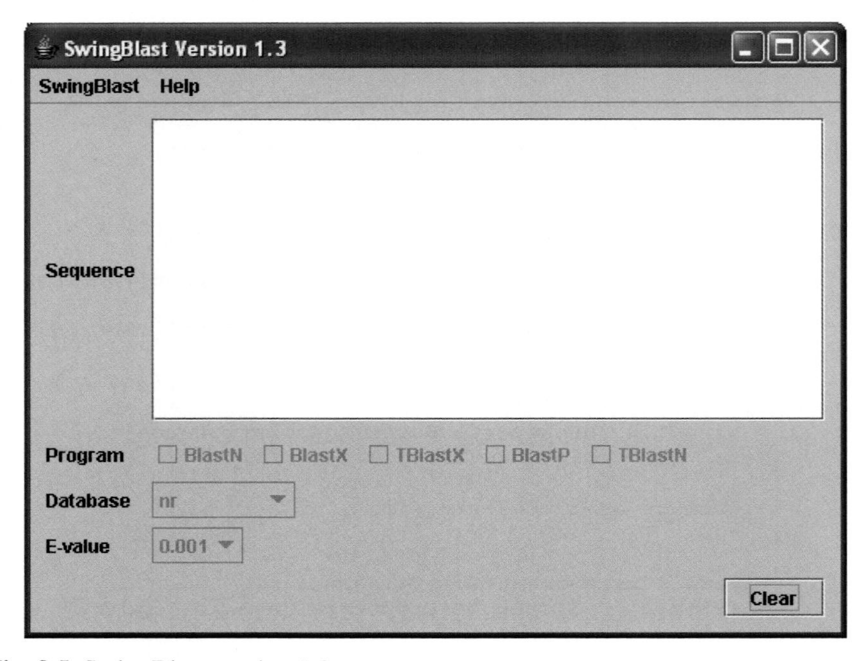

Fig. 3.5. SwingBlast version 1.3

We will enhance `SwingBlast` in a number of ways in this Chapter. In particular, we will:

1. Introduce a Format button to convert the entered sequence into Fasta format. In the earlier version, the `SwingBlast` application required the user to lose focus away from the text area in order to perform the formatting.

2. Add a Submit button to send sequences for BLAST searches.

3. Add code behind the BLAST programs (BLASTN, BLASTX, etc.) so that checking the boxes will enable the user to run the corresponding BLAST programs.

4. Add functionality to prompt the user to save BLAST search results.

We will call the resulting application `SwingBlast` Version 2.1. We add the button widgets we need for the `SwingBlast` application as we did previously.

```
private JButton formatBtn;
formatBtn = new JButton("Format Sequence");
```

To place the button in the GUI, we use the `JPanel` object:

```
JPanel panel = new JPanel();
panel.add(formatBtn);
seqPanel.add(panel, BorderLayout.CENTER);
```

To format a sequence, we first need to know when the text area is populated with a sequence. To do this we implement an event listener, which was explained in Chapter 2.

```
    private void addListeners() {
      formatBtn.addActionListener(new ActionListener() {
        public void actionPerformed(ActionEvent e) {
          // Check if sequence is DNA, RNA or protein
          // Retrieve text entered in the text area
          String sequenceText = sequenceArea.getText();
          if (sequenceText == null || sequenceText.length()
== 0) {
            cleanAllParameters();
            return;
          }
        }
      }
    }
```

The `cleanAllParameters()` method clears the text in the text area and disables the `enableFunctions()` method which checks the entered sequence for type, that is, DNA, RNA or protein.

```
    private void cleanAllParameters() {
      sequenceArea.setText("");
      enableFunctions(-1);
    }
```

Next, let's add the code to format the input sequence. We will program the format button to cause the sequence in the text area to be wrapped into lines of 50 bases each and add a Fasta header at the top using the code below:

```
      private StringBuffer format(String sequence) {
          int i = 1;
          final int seqLen = sequence.length();
          StringBuffer sb = new StringBuffer(seqLen);
          if (seqLen > 50) {
              char[] chars = sequence.toCharArray();
```

```
                for (int j = 0; j < chars.length; j++) {
                    sb.append(chars[j]);

                    if (i++ % 50 == 0) {
                        sb.append("\n");
                    }
                }
            } else {
                sb.append(sequence);
            }
            return sb;
    }
```

We had described the logic to program the check boxes for the various BLAST algorithms based on the input sequence earlier in Chapter 2. The application at this stage appears as is shown in **Fig. 3.6**. Let's test the application with a fragment of the human cystic fibrosis transmembrane conductance regulator (CFTR) mRNA sequence (gi: 90421312) we had described in Chapter 2. Compile and run the application and paste the sequence in the text area (**Fig. 3.7**).

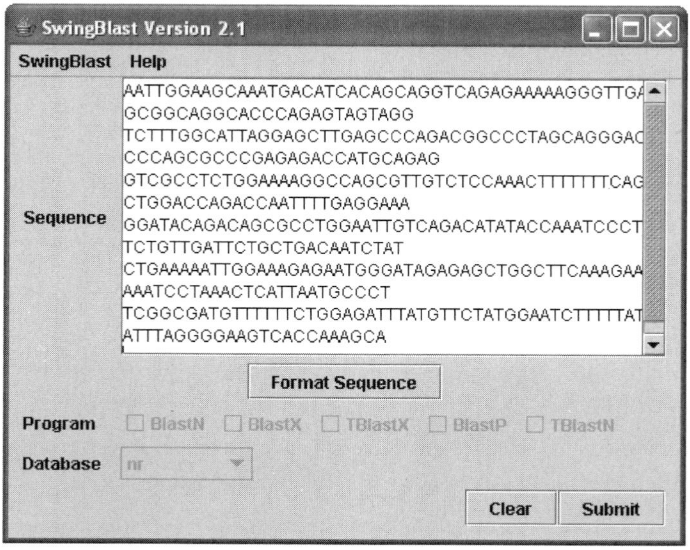

Fig. 3.6. SwingBlast Version 2.1

The formatted sequence is shown below (**Fig. 3.7**).

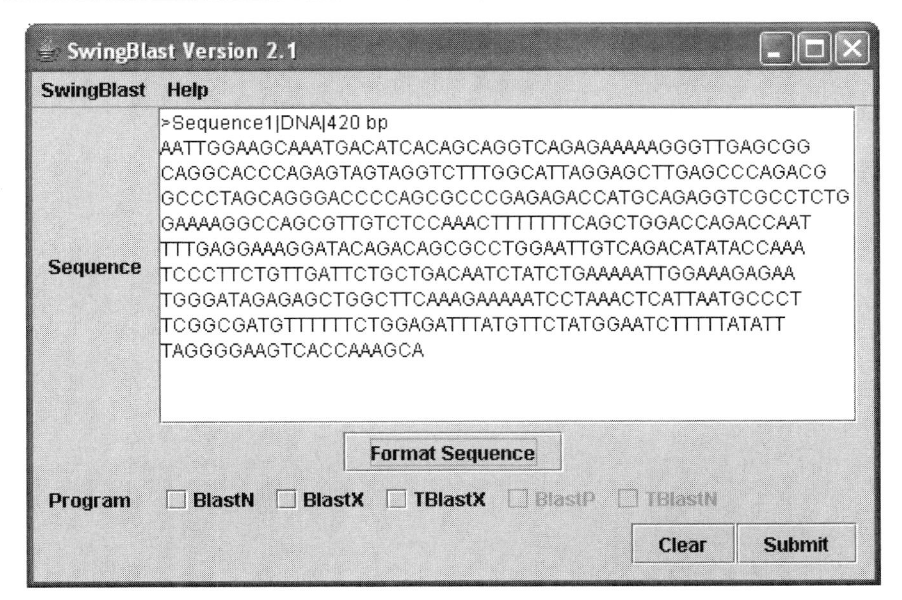

Fig. 3.7. Fasta formatted DNA sequence

To align the Fasta format sequence properly, we had described the use of a monospaced font earlier for the DNA alphabet:

```
        final Font sf = sequenceArea.getFont();
        Font f = new Font("Monospaced", sf.getStyle(), sf.getSize());
        sequenceArea.setFont(f);
```

An explicit monospace font such as *Courier* can also be used provided it is installed on your machine. The application with the sequence formatted in monospace font is shown in **Fig. 3.8**.

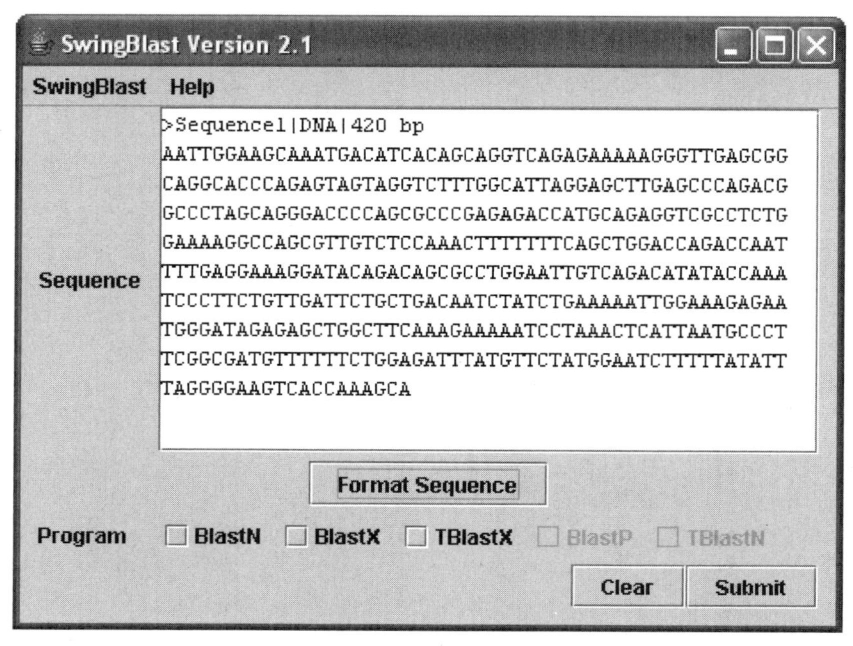

Fig. 3.8. Fasta formatting with a monospace font

Note that the application first checks if the sequence is in Fasta format before applying the formatting. If a sequence that is pasted is already in Fasta format, clicking the "`Format Sequence`" button does not have any effect. The user can now select one or more of the available BLAST options and hit `Submit` to run the search. Let's run a search with the partial CFTR sequence using BLASTN and BLASTX using SwingBlast 2.1 (**Fig. 3.9**).

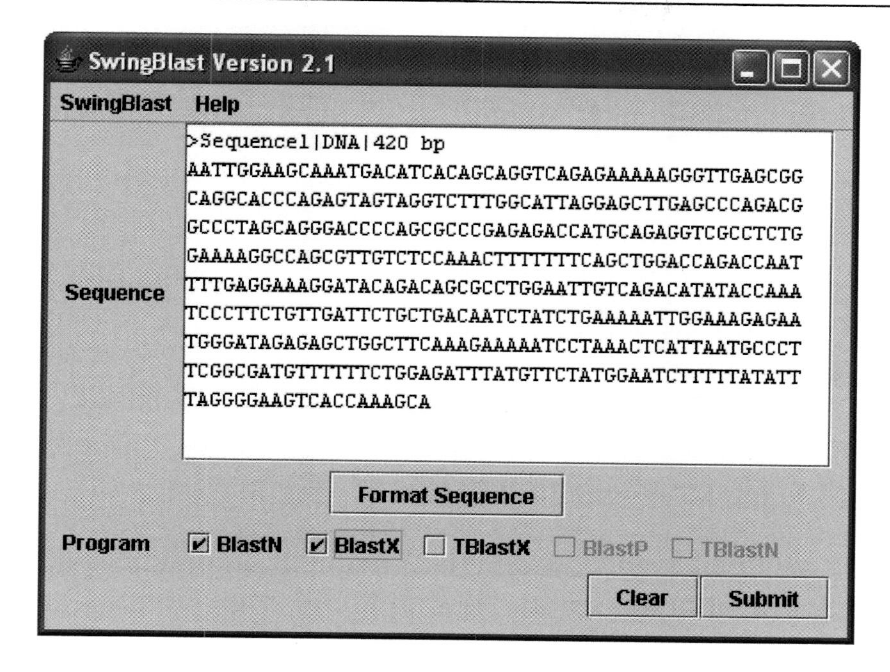

Fig. 3.9. Running a BLASTN and BLASTX search

We get a notification once each of the requested BLAST search is complete as shown below for the BLASTN search (**Fig 3.10**). After each analysis is complete, the application also prompts the user to save the results of the search in a local text file (**Fig. 3.11 - 3.12**).

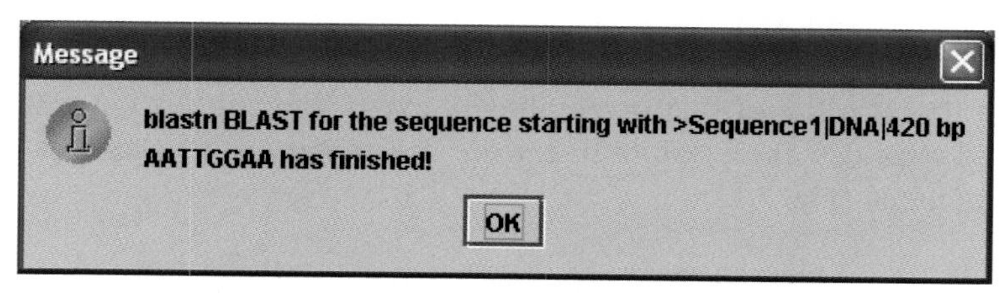

Fig. 3.10. BLAST search status notification

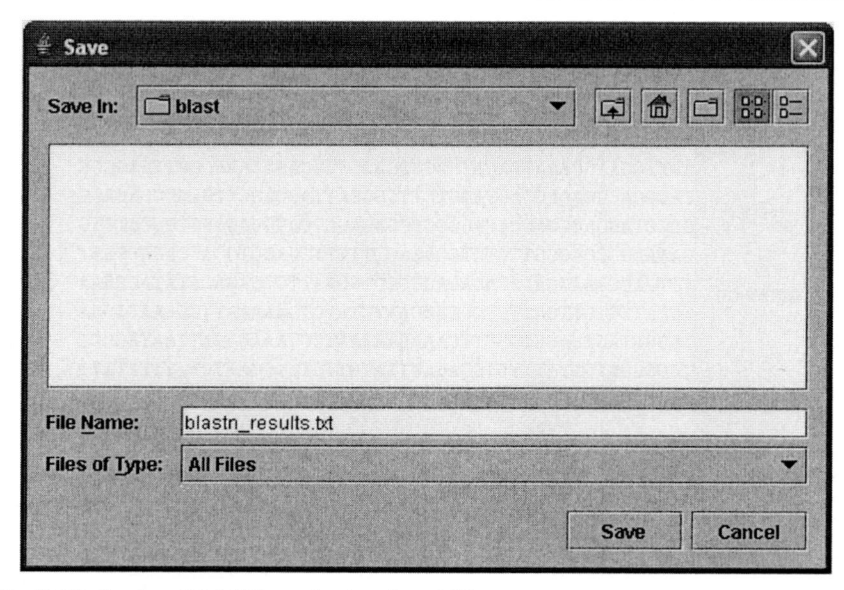

Fig. 3.11. Saving BLAST results in a local file

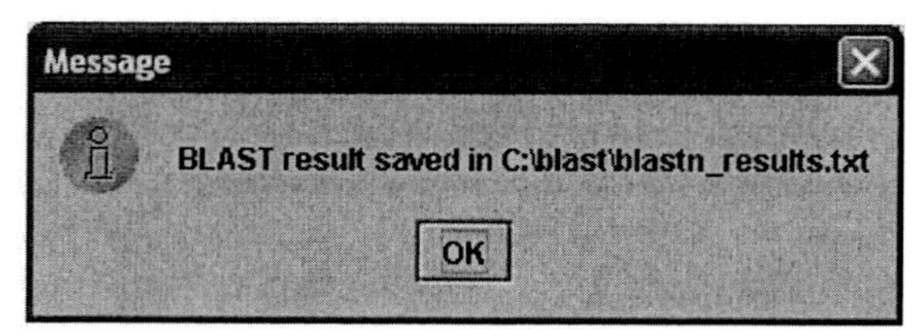

Fig. 3.12. Saving BLAST results in a local file

Note that if a file of that name already exists, the application warns the user and provides an option to overwrite the existing file or save it with a different name (**Fig. 3.13**).

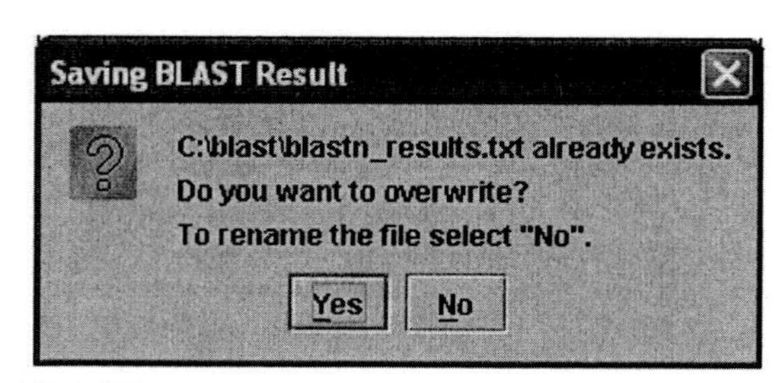

Fig. 3.13. Saving BLAST results in a different file

This functionality is implemented within the `saveBlast()` function as shown in **Listing 3.5**.

Listing 3.5. The saveBlast() function

```
    private void saveBlast(String tmpFileName) {
      final String fileNameFromUser = getFileNameFromUser();
      if (fileNameFromUser == null)
        return;

      final File tmpFile = new File(tmpFileName);
      final File userFile = new File(fileNameFromUser);
      if (userFile.exists()) {
        String   errMes   =   fileNameFromUser   +   "   already
exists.\nDo you want to overwrite?\n" + "To rename the file
select \"No\".";
        int   choice   =   JOptionPane.showConfirmDialog(this,
errMes, "Saving BLAST Result", JOptionPane.YES_NO_OPTION);
        if (choice == JOptionPane.YES_OPTION) {
          userFile.delete();
          tmpFile.renameTo(userFile);
        } else {
          saveBlast(tmpFileName);
        }
      } else {
        tmpFile.delete();
        JOptionPane.showMessageDialog(SequenceForm2_2.this,
"BLAST result saved in " + fileNameFromUser);
      }
    }
```

If the user doesn't want to overwrite an existing file, a new file name must be supplied. This is implemented in the `getFileNameFromUser()` function described below (**Listing 3.6**).

Listing 3.6. The getFileNameFromUser() function

```
    private String getFileNameFromUser() {
      JFileChooser fc = new JFileChooser();
      if              (fc.showSaveDialog(this)               ==
JFileChooser.APPROVE_OPTION) {
        return fc.getSelectedFile().getAbsolutePath();
      } else {
        return null;
      }
    }
```

The BLAST results can be viewed in their raw format (as saved in the text file above) using a text editor (**Fig. 3.14**) for parsing to diplay the results in a graphical format. The complete code for `SwingBlast` version 2.1 is shown in **Listing 3.7**.

Fig. 3.14. Viewing saved BLAST results in text format

Listing 3.7. SwingBlast Version 2.1

```
package org.jfb.swingblast2;

import org.apache.regexp.RE;
import org.apache.regexp.RESyntaxException;
import org.jfb.blast.Blast;
import org.jfb.blast.BlastException;
import org.jfb.blast.BlastManager;
import org.jfb.jqblast.RequestIdentifier;

import javax.swing.*;
import java.awt.*;
import java.awt.event.ActionEvent;
import java.awt.event.ActionListener;
import java.io.File;
import java.util.ArrayList;
import java.util.HashMap;
import java.util.Observable;
import java.util.Observer;

public class SwingBlast2_1 extends JFrame {
```

```
      private static final String APP_NAME = "SwingBlast";
      private static final String APP_VERSION = "Version
2.1";

      private static final Dimension LABEL_PREFERRED_SIZE =
new Dimension(57, 16);
      private static final Dimension COMBO_PREFERRED_SIZE =
new Dimension(60, 25);
      private static final Dimension CP_PREF_SIZE = new
Dimension(480, 380);

      private static final int TYPE_DNA = 0;
      private static final int TYPE_RNA = 1;
      private static final int TYPE_PROTEIN = 2;

      private static final String[] BLAST_PROGRAMS_DNA = new
String[]{"BlastN", "BlastX", "TBlastX"};
      private static final String[] BLAST_PROGRAMS_PROTEIN =
new String[]{"BlastP", "TBlastN"};
      private static final String[] DATABASES = new
String[]{"nr", "est_human"};
      private static final String[] EVALUES = new
String[]{"0.001", "0.01", "0.1", "1", "10", "100"};

      private JComponent newContentPane;
      private JTextArea sequenceArea;
      private JScrollPane scrollPaneArea;

      private JCheckBox[] chbDna;
      private JCheckBox[] chbProtein;
      private JComboBox cobDbs;
      private JComboBox cobEvalues;

      private JButton submitBtn;
      private JButton formatBtn;
      private JButton clearBtn;

      private JMenuItem aboutItem;
      private JMenuItem quitItem;
      private static final double SEQ_THRESHOLD = 0.85;
      private static final int TYPE_UNKNOWN = -1;
      private int typeOfSequence;
      private static final String SEQ_HEADER_GEN =
">Sequence1|";
      private static final int SUB_MAX = 30;

      static {
          try {
              Class.forName("org.jfb.jqblast.JQBlast");
          } catch (ClassNotFoundException e) {
              e.printStackTrace();
          }
      }
```

```
        public SwingBlast2_1() {
            super();
        }

        private void seqFormInit() {
            setTitle(APP_NAME + " " + APP_VERSION);
            setDefaultCloseOperation(JFrame.EXIT_ON_CLOSE);
            newContentPane = new JPanel();
            newContentPane.setOpaque(true);
            newContentPane.setLayout(new BorderLayout());

            setContentPane(newContentPane);

            // Add the menu bar.
            JMenuBar menu = new JMenuBar();
            JMenu swingBlastMenu = new JMenu(APP_NAME);
            quitItem = new JMenuItem("Quit");
            swingBlastMenu.add(quitItem);
            menu.add(swingBlastMenu);

            JMenu helpMenu = new JMenu("Help");
            aboutItem = new JMenuItem("About");
            helpMenu.add(aboutItem);
            menu.add(helpMenu);
            setJMenuBar(menu);

            // Create the seqLbl pane
            JPanel sequencePanel = new JPanel();
            JLabel seqLbl = new JLabel("Sequence");
            sequenceArea = new JTextArea();
            sequenceArea.setLineWrap(true);
            final Font sf = sequenceArea.getFont();
            Font  f  =  new  Font("Monospaced",  sf.getStyle(),
sf.getSize());
            sequenceArea.setFont(f);
            scrollPaneArea = new JScrollPane(sequenceArea);
            scrollPaneArea.setPreferredSize(new  Dimension(300,
200));
            formatBtn = new JButton("Format Sequence");

            sequencePanel.setLayout(new
BoxLayout(sequencePanel, BoxLayout.LINE_AXIS));
            sequencePanel.add(seqLbl);
            sequencePanel.add(Box.createRigidArea(new
Dimension(10, 0)));
            sequencePanel.add(scrollPaneArea);

            JPanel seqPanel = new JPanel();
            seqPanel.setLayout(new BorderLayout());
            seqPanel.add(sequencePanel, BorderLayout.NORTH);
            JPanel panel = new JPanel();
            panel.add(formatBtn);
```

```
        seqPanel.add(panel, BorderLayout.CENTER);

        //Lay out the buttons from left to right.
        JPanel buttonPane = new JPanel();
        submitBtn = new JButton("Submit");
        clearBtn = new JButton("Clear");

        buttonPane.setLayout(new      BoxLayout(buttonPane,
BoxLayout.LINE_AXIS));
        buttonPane.add(Box.createHorizontalGlue());
        buttonPane.add(Box.createRigidArea(new
Dimension(10, 0)));
        buttonPane.add(clearBtn);
        buttonPane.add(submitBtn);

        JPanel jPanel = new JPanel();
        jPanel.setLayout(new BorderLayout());
        jPanel.setBorder(BorderFactory.createEmptyBorder(0,
10, 10, 10));
        jPanel.add(seqPanel, BorderLayout.NORTH);
        jPanel.add(createProgramPanel(),
BorderLayout.CENTER);
        jPanel.add(buttonPane, BorderLayout.SOUTH);

        newContentPane.add(jPanel, BorderLayout.CENTER);
        newContentPane.setPreferredSize(CP_PREF_SIZE);

        //Display the window.
        pack();
        Dimension               screenSize              =
Toolkit.getDefaultToolkit().getScreenSize();
        setLocation((screenSize.width - CP_PREF_SIZE.width)
/ 2,
                (screenSize.height - CP_PREF_SIZE.height) /
2);
        setVisible(true);
        addListeners();
    }

    private JPanel createProgramPanel() {
        // Let's get the program panel using the same
layout
        JPanel programPanel = new JPanel();
        JLabel program = new JLabel("Program");
        program.setPreferredSize(LABEL_PREFERRED_SIZE);
        chbDna = new JCheckBox[BLAST_PROGRAMS_DNA.length];
        String blastProgram;
        for (int i = 0; i < BLAST_PROGRAMS_DNA.length; i++)
{
            blastProgram = BLAST_PROGRAMS_DNA[i];
            chbDna[i] = new JCheckBox(blastProgram);
            chbDna[i].setMaximumSize(COMBO_PREFERRED_SIZE);
        }
```

```
          chbProtein                    =                    new
JCheckBox[BLAST_PROGRAMS_PROTEIN.length];
          for (int i = 0; i < BLAST_PROGRAMS_PROTEIN.length;
i++) {
              blastProgram = BLAST_PROGRAMS_PROTEIN[i];
              chbProtein[i] = new JCheckBox(blastProgram);

chbProtein[i].setMaximumSize(COMBO_PREFERRED_SIZE);
          }

          programPanel.setLayout(new  BoxLayout(programPanel,
BoxLayout.LINE_AXIS));
          programPanel.add(program);
          programPanel.add(Box.createRigidArea(new
Dimension(10, 0)));
          for (int i = 0; i < chbDna.length; i++) {
              programPanel.add(chbDna[i]);
              programPanel.add(Box.createRigidArea(new
Dimension(5, 0)));
          }
          for (int i = 0; i < chbProtein.length; i++) {
              programPanel.add(chbProtein[i]);
              if (i + 1 < chbProtein.length)
                  programPanel.add(Box.createRigidArea(new
Dimension(5, 0)));
          }
          programPanel.add(Box.createHorizontalGlue());
          JPanel paramPanel = new JPanel();
          paramPanel.setLayout(new      BoxLayout(paramPanel,
BoxLayout.PAGE_AXIS));

          paramPanel.add(programPanel);
          paramPanel.add(Box.createRigidArea(new Dimension(0,
5)));

          // Create the database panel using the same layout
          JPanel databasePanel = new JPanel();
          JLabel database = new JLabel("Database");
          database.setPreferredSize(LABEL_PREFERRED_SIZE);
          cobDbs = new JComboBox(DATABASES);
          cobDbs.setMaximumSize(COMBO_PREFERRED_SIZE);

          databasePanel.setLayout(new
BoxLayout(databasePanel, BoxLayout.LINE_AXIS));
          databasePanel.add(database);
          databasePanel.add(Box.createRigidArea(new
Dimension(10, 0)));
          databasePanel.add(cobDbs);
          databasePanel.add(Box.createHorizontalGlue());
          paramPanel.add(databasePanel);
          paramPanel.add(Box.createRigidArea(new Dimension(0,
5)));
```

```
        // Create the E-Value panel using the same layout
        JPanel evaluePanel = new JPanel();
        JLabel eValue = new JLabel("E-value");
        eValue.setPreferredSize(LABEL_PREFERRED_SIZE);
        cobEvalues = new JComboBox(EVALUES);
        cobEvalues.setMaximumSize(COMBO_PREFERRED_SIZE);

        evaluePanel.setLayout(new    BoxLayout(evaluePanel,
BoxLayout.LINE_AXIS));
        evaluePanel.add(eValue);
        evaluePanel.add(Box.createRigidArea(new
Dimension(10, 0)));
        evaluePanel.add(cobEvalues);
        evaluePanel.add(Box.createHorizontalGlue());
        paramPanel.add(evaluePanel);
        paramPanel.add(Box.createRigidArea(new Dimension(0,
5)));


        enableFunctions(TYPE_UNKNOWN);
        return paramPanel;
    }

    private void addListeners() {
        quitItem.addActionListener(new ActionListener() {
            public void actionPerformed(ActionEvent e) {
                System.exit(0);
            }
        });

        aboutItem.addActionListener(new ActionListener() {
            public void actionPerformed(ActionEvent e) {

JOptionPane.showMessageDialog(SwingBlast2_1.this, APP_NAME +
" " + APP_VERSION,
                            "About     "     +     APP_NAME,
JOptionPane.INFORMATION_MESSAGE);
            }
        });

        submitBtn.addActionListener(new ActionListener() {
            public void actionPerformed(ActionEvent e) {
                StringBuffer     errMes     =     new
StringBuffer("<HTML>Please    provide    the    following
parameters:<BR>");
                String sequence = sequenceArea.getText();
                boolean misPar = false;
                if (sequence == null || sequence.length()
== 0) {
                    errMes.append("- Sequence<BR>");
                    misPar = true;
                }
```

```
                    String      database      =      (String)
cobDbs.getSelectedItem();
                    String[] blastTypes = getBlastTypes();
                    if (blastTypes == null || blastTypes.length
== 0) {
                        errMes.append("- blast<BR>");
                        misPar = true;
                    }
                    final String endOfPleaseMes = "</html>";
                    errMes.append(endOfPleaseMes);
                    if (misPar) {

JOptionPane.showMessageDialog(SwingBlast2_1.this, errMes);
                        return;
                    }
                    String       evalue       =       (String)
cobEvalues.getSelectedItem();
                    runBlasts(sequence,  blastTypes,  database,
evalue);
                }
            });

            clearBtn.addActionListener(new ActionListener() {
                public void actionPerformed(ActionEvent e) {
                    cleanAllParameters();
                }
            });

            formatBtn.addActionListener(new ActionListener() {
                public void actionPerformed(ActionEvent e) {
                    // Check sequence type
                    // Retrieve text entered in the text area
                    final      String      sequenceText      =
sequenceArea.getText();
                    if    (sequenceText    ==    null    ||
sequenceText.length() == 0) {
                        cleanAllParameters();
                        return;
                    }
                    // Format sequence in FASTA format
                    int idx = sequenceText.indexOf(">");
                    final boolean fastaFormatted = idx != -1;
                    String header = null;
                    String sequence = "";

                    if (fastaFormatted) {
                        int             returnIdx             =
sequenceText.indexOf("\n");

                        if (returnIdx != -1) {
                            header = sequenceText.substring(0,
returnIdx);
                            sequence                          =
```

```
sequenceText.substring(returnIdx                    +              1,
sequenceText.length()).replaceAll("\\s", "").toLowerCase();
                        }
                        // Check if sequence entered
                        updateSequenceArea(header,     sequence,
fastaFormatted);
                    } else {
                        updateSequenceArea(SEQ_HEADER_GEN,
sequenceText.toLowerCase(), fastaFormatted);
                    }
                }
            });
        }

        private  void  updateSequenceArea(String  header,  String
sequence, boolean fastaFormatted) {
            String seqText;
            if (sequence.length() == 0)
                return;

            // Retrieve sequence type
            this.typeOfSequence = TYPE_UNKNOWN;
            try {
                this.typeOfSequence                                =
getSequenceType(sequence);
            } catch (RESyntaxException rese) {
                rese.printStackTrace();
            }

            String type = null;
            String unitOfLength = null;

            switch (this.typeOfSequence) {
                case TYPE_DNA:
                    type = "DNA";
                    unitOfLength = " bp";
                    break;
                case TYPE_RNA:
                    type = "RNA";
                    unitOfLength = " bp";
                    break;
                case TYPE_PROTEIN:
                    type = "Protein";
                    unitOfLength = " aa";
                    break;
                default:
                    type = "N/A";
                    unitOfLength = " N/A";
            }

            if (!fastaFormatted) {
                seqText    =    header    +    type    +    "|"    +
sequence.length()     +     unitOfLength     +     "\n"     +
```

```
format(sequence.toUpperCase());
          } else {
              seqText     =     header     +     "\n"     +
format(sequence.toUpperCase());
          }

          // Display results in the sequence area
          sequenceArea.setText(seqText);

          enableFunctions(this.typeOfSequence);
      }

      private StringBuffer format(String seq) {
          int i = 1;
          String sequence = seq.replaceAll("\n", "");
          final int seqLen = sequence.length();
          StringBuffer sb = new StringBuffer(seqLen);
          if (seqLen > 50) {
              char[] chars = sequence.toCharArray();
              for (int j = 0; j < chars.length; j++) {
                  sb.append(chars[j]);

                  if (i++ % 50 == 0) {
                      sb.append("\n");
                  }
              }
          } else {
              sb.append(sequence);
          }
          return sb;
      }

      private void runBlasts(final String sequence, String[]
blastTypes, String database, String evalue) {
          Map param = new HashMap();
          param.put("sequenceText", sequence);
          param.put("database", database);
          param.put("eValue", evalue);

          final Observer observer = new Observer() {
              public void update(Observable o, Object arg) {
                  System.out.println("" + arg);
              }
          };

          try {
              for (int i = 0; i < blastTypes.length; i++) {
                  final String blastType = blastTypes[i];
                  final Map tmp = new HashMap(param);
                  tmp.put("blastType", blastType);
                  Thread t = new Thread(new Runnable() {
                      public void run() {
                          try {
```

```
                              final    Blast    blast    =
BlastManager.createBlast();
                              blast.addObserver(observer);
                              RequestIdentifier
requestIdentifier            =          (RequestIdentifier)
blast.submitQuery(tmp);
                              final   String   fileName   =
blast.requestResult(requestIdentifier).toString();
                              final StringBuffer sb =
                                    new
StringBuffer().append(blastType).append("  BLAST  for  the
sequence starting with ")

.append(sequence.length() > SUB_MAX ? sequence.substring(0,
SUB_MAX) : sequence).append(" has finished!");
                              Runnable   runnable   =   new
Runnable() {
                                 public void run() {

JOptionPane.showMessageDialog(SwingBlast2_1.this,
sb.toString());
                                     saveBlast(fileName);
                                 }
                              };

SwingUtilities.invokeLater(runnable);
                          } catch (BlastException be) {
                              be.printStackTrace();
                          } catch (Throwable e) {
                              e.printStackTrace();
                          }
                      }
                  });
                  t.start();
              }
          } catch (Throwable e) {
              e.printStackTrace();
          }
      }

      private void saveBlast(String tmpFileName) {
          final        String        fileNameFromUser        =
getFileNameFromUser();
          if (fileNameFromUser == null)
              return;

          final File tmpFile = new File(tmpFileName);
          final File userFile = new File(fileNameFromUser);
          String finalName = tmpFileName;
          if (userFile.exists()) {
              String errMes = fileNameFromUser + " already
exists.\nDo you want to overwrite?.";
              int                    choice                =
```

```
JOptionPane.showConfirmDialog(this,  errMes,  "Saving  BLAST
Result", JOptionPane.YES_NO_OPTION);
            if (choice == JOptionPane.YES_OPTION) {
                boolean              renamed              =
tmpFile.renameTo(userFile);
                if (renamed) {
                    tmpFile.delete();
                    finalName = fileNameFromUser;
                }
            } else {
                saveBlast(tmpFileName);
                return;
            }
        } else {
            boolean renamed = tmpFile.renameTo(userFile);
            if (renamed) {
                tmpFile.delete();
                finalName = fileNameFromUser;
            }
        }
        JOptionPane.showMessageDialog(SwingBlast2_1.this,
"BLAST result saved in " + finalName);
    }

    private String getFileNameFromUser() {
        JFileChooser fc = new JFileChooser();
        if          (fc.showSaveDialog(this)              ==
JFileChooser.APPROVE_OPTION) {
            return fc.getSelectedFile().getAbsolutePath();
        } else {
            return null;
        }
    }

    protected void finalize() throws Throwable {
        super.finalize();

    }

    private void cleanAllParameters() {
        sequenceArea.setText("");
        enableFunctions(-1);
    }

    private String[] getBlastTypes() {
        JCheckBox[] allTypes  =  typeOfSequence  ==  TYPE_DNA
|| typeOfSequence == TYPE_RNA
                ? chbDna : typeOfSequence == TYPE_PROTEIN ?
chbProtein : null;
        if (allTypes == null) return null;

        ArrayList types = new ArrayList();
        for (int i = 0; i < allTypes.length; i++) {
```

```
                JCheckBox cb = allTypes[i];
                if (cb.isSelected())
                    types.add(cb.getText().toLowerCase());
            }
            final String[] res = new String[types.size()];
            types.toArray(res);
            return res;
        }

        private void enableFunctions(int typeOfSequence) {
            if (typeOfSequence == TYPE_DNA || typeOfSequence ==
TYPE_RNA) {
                setChb(chbDna, true);
                setChb(chbProtein, false);
                setCob(cobDbs, true);
                setCob(cobEvalues, true);
            } else if (typeOfSequence == TYPE_PROTEIN) {
                setChb(chbProtein, true);
                setChb(chbDna, false);
                setCob(cobDbs, true);
                setCob(cobEvalues, true);
            } else {
                setChb(chbProtein, false);
                setChb(chbDna, false);
                setCob(cobDbs, false);
                setCob(cobEvalues, false);
            }
        }

        private static void setChb(JCheckBox[] boxes, boolean
value) {
            for (int i = 0; i < boxes.length; i++) {
                boxes[i].setEnabled(value);
                boxes[i].setSelected(false);
            }
        }

        private static void setCob(JComboBox component, boolean
value) {
            component.setEnabled(value);
            component.setSelectedIndex(0);
        }

        public static int getSequenceType(String sequence)
throws RESyntaxException {
            RE re = new RE("[actgnACGTN]+");
            String[] strings = re.split(sequence);
            int numbOfLettersOtherThanATGCNs = 0;

            for (int i = 0; i < strings.length; i++) {
                numbOfLettersOtherThanATGCNs                    +=
strings[i].length();
            }
```

```
            int length = sequence.length();
            int        numbOfACGTNs        =        length        -
numbOfLettersOtherThanATGCNs;

            re = new RE("[uU]+");
            strings = re.split(sequence);
            int numbOfLettersOtherThanUs = 0;

            for (int i = 0; i < strings.length; i++) {
                numbOfLettersOtherThanUs                           +=
strings[i].length();
            }
            int      numbOfUs       =       sequence.length()       -
numbOfLettersOtherThanUs;

            if (numbOfACGTNs / (double) length > SEQ_THRESHOLD)
{
                return TYPE_DNA;
            } else if ((numbOfACGTNs + numbOfUs) / (double)
length > SEQ_THRESHOLD) {
                return TYPE_RNA;
            } else {
                return TYPE_PROTEIN;
            }
        }

        public static void main(String[] args) {
            SwingUtilities.invokeLater(new Runnable() {
                public void run() {
                    final  SwingBlast2_1  sequenceForm  =  new
SwingBlast2_1();
                    sequenceForm.seqFormInit();
                }
            });
        }
    }
```

Retrieving Sequences From GenBank Using BioJava

Frequently, users know GI numbers of sequences that they use regularly in their research and it is normal for them to submit a GI number of the corresponding sequence for BLAST searches on the NCBI BLAST service. We will next implement a feature in `SwingBlast` whereby users can retrieve a sequence from GenBank based on its GI number. We will use existing *BioJava* routines to retrieve sequences corresponding to a GenBank ID that users may enter into the sequence field. We will need the following BioJava libraries to accomplish this task:

```
org.biojava.bio.seq.Sequence;

org.biojava.bio.seq.db.GenbankSequenceDB;

org.biojava.bio.seq.io.SeqIOTools;
```

These libraries can be obtained from the BioJava website (Binary for J2SE 1.4 or later, as of this writing) at the following URL:

`http://biojava.org/wiki/BioJava:Download`

Since users have the option of entering sequences directly into the sequence field, we need to first test if the entered text is a sequence or a genbank ID. We will do this using *regular expressions* as outlined below:

```
text = text.replaceAll("\\s", "");
RE re = null;
try {
    re = new RE("[0-9]+");
} catch (RESyntaxException e1) {
    e1.printStackTrace();
}

boolean isGenBankID = re.match(text);
```

We then create a new instance of the class `GenbankSequenceDB` that will retrieve the Genbank record. `seqObject` contains the entire GenBank record, that is, the header information, any sequence features and annotation and the actual nucleotide or amino acid sequence.

```
seqObject = genbankSequenceDB.getSequence(text);
```

To see the content of the sequence object retrieved we can write it to the system output using `SeqIOTools` as followed:

```
SeqIOTools.writeGenbank(System.out, seqObject);
```

To grab only the sequence we then use the method `seqString()` from the `seqObject`.

```
sequence = seqObject.seqString();
```

The complete code is as follows:

```
import org.biojava.bio.seq.Sequence;
import org.biojava.bio.seq.db.GenbankSequenceDB;
```

```
import org.biojava.bio.seq.io.SeqIOTools;

text = text.replaceAll("\\s", "");
RE re = null;
try {
    re = new RE("[0-9]+");
} catch (RESyntaxException e1) {
    e1.printStackTrace();
}

boolean isGenBankID = re.match(text);

if (isGenBankID) {
    GenbankSequenceDB        genbankSequenceDB        =        new
GenbankSequenceDB();
    header = "GI:" + text;
    Sequence seqObject = null;
    try {
        seqObject = genbankSequenceDB.getSequence(text);
        SeqIOTools.writeGenbank(System.out, seqObject);
    } catch (Exception e) {
        e.printStackTrace();
    }
    sequence = seqObject.seqString();
}
```

The "Format Sequence" in the application will now have a dual function when a GI number is pasted in the text area – it will retrieve the sequence from GenBank and simultaneously convert it into the Fasta format. We will call this version of the application `SwingBlast` version 2.2. The code for SwingBlast Version 2.2 with this feature implemented is shown in **Listing 3.8**.

Listing 3.8. SwingBlast Version 2.2

```
Runnable runnable = new Runnable() {
    public void run() {
        String seq = null;
        final          boolean          isGenBankID          =
GenbankDB.isGenBankId(sequenceText);

        if (isGenBankID) {
            boolean canGetSeq = true;
            GenbankSequenceDB    genbankSequenceDB    =    new
GenbankSequenceDB();
            header = "GI:" + text;
            Sequence seqObject = null;
            try {
```

```
                    seqObject                                    =
genbankSequenceDB.getSequence(text);
                    SeqIOTools.writeGenbank(System.out,
seqObject);
                } catch (Exception e) {
                    e.printStackTrace();
                }
                seq = seqObject.seqString();
                if  (seq  ==  null  ||  seq.length()  ==  0  ||
!canGetSeq) {

JOptionPane.showMessageDialog(SwingBlast2_2.this,
                        "Cannot get the sequence for GenBank ID
" + sequenceText);
                return;
            }
        }

        SwingBlast2_2.this.sequence = seq;
        Runnable runnableAwt = new Runnable() {
            public void run() {
                String seqFin = SwingBlast2_2.this.sequence;
                String header = null;
                String sequence = "";

                if (isGenBankID) {
                    int i = seqFin.indexOf("\n");
                    header = seqFin.substring(0, i);
                    sequence     =     seqFin.substring(i     +
"\n".length(), seqFin.length());
                } else {
                    sequence = sequenceText.toLowerCase();
                    header = SEQ_HEADER_GEN;
                }
                // We first check that there is something.
                updateSequenceArea(header,            sequence,
fastaFormatted, isGenBankID);
            }
        };
        SwingUtilities.invokeLater(runnableAwt);
    };
    new Thread(runnable).start();
```

Fig. 3.15 and **Fig. 3.16** below show the results of pasting a GenBank Id in the sequence area of SwingBlast Version 2.2.

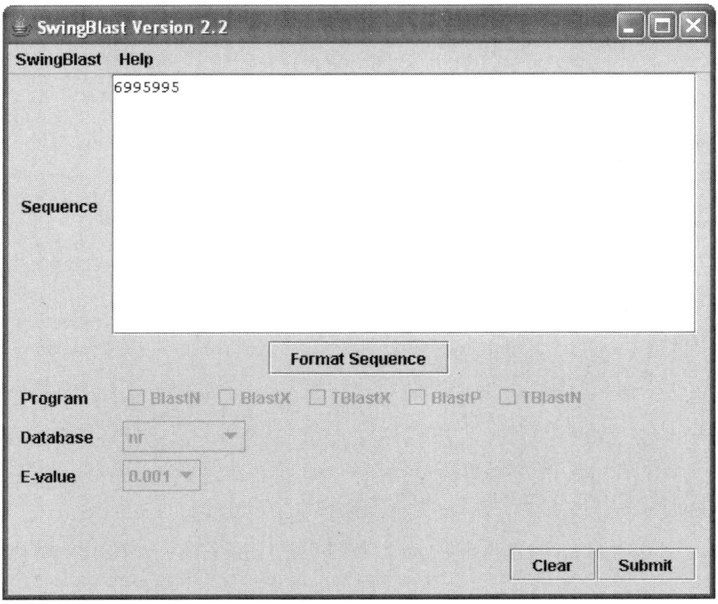

SwingBlast Version 2.2

SwingBlast Help

Sequence

6995995

Format Sequence

Program BlastN BlastX TBlastX BlastP TBlastN

Database nr

E-value 0.001

Clear Submit

Fig. 3.15. Pasting GI number in the text area for sequence retrieval

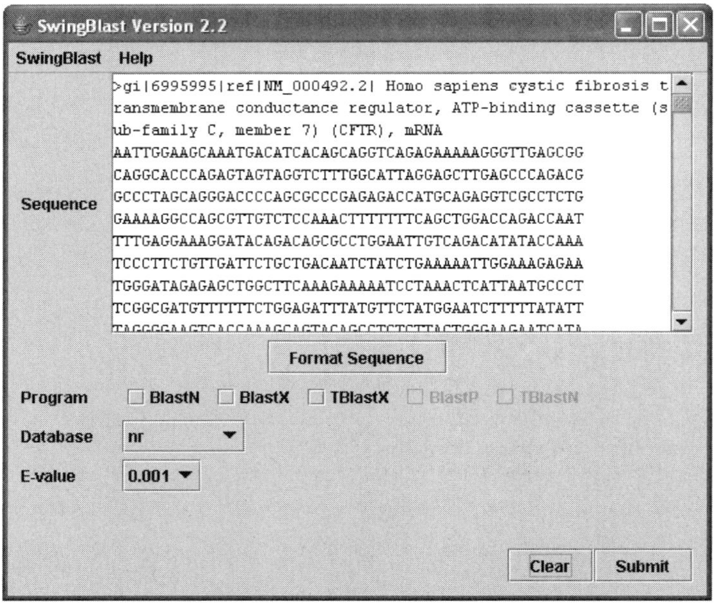

SwingBlast Version 2.2

SwingBlast Help

Sequence

```
>gi|6995995|ref|NM_000492.2| Homo sapiens cystic fibrosis t
ransmembrane conductance regulator, ATP-binding cassette (s
ub-family C, member 7) (CFTR), mRNA
AATTGGAAGCAAATGACATCACAGCAGGTCAGAGAAAAAGGGTTGAGCGG
CAGGCACCCAGAGTAGTAGGTCTTTGGCATTAGGAGCTTGAGCCCAGACG
GCCCTAGCAGGGACCCCAGCGCCCGAGAGACCATGCAGAGGTCGCCTCTG
GAAAAGGCCAGCGTTGTCTCCAAACTTTTTTTCAGCTGGACCAGACCAAT
TTTGAGGAAAGGATACAGACAGCGCCTGGAATTGTCAGACATATACCAAA
TCCCTTCTGTTGATTCTGCTGACAATCTATCTGAAAAATTGGAAAGAGAA
TGGGATAGAGAGCTGGCTTCAAAGAAAAATCCTAAACTCATTAATGCCCT
TCGGCGATGTTTTTTCTGGAGATTTATGTTCTATGGAATCTTTTTATATT
```

Format Sequence

Program BlastN BlastX TBlastX BlastP TBlastN

Database nr

E-value 0.001

Clear Submit

Fig. 3.16. Retrieving a sequence from GenBank from its GI number

Although the Fasta header in **Fig. 3.16** appears to run over multiple lines, it is actually a single line that has wrapped over because of the size of the text area.

Retrieving GenBank Without BioJava

This is how one would implement the retrieval of the sequence using GenBank ID and NCBI web application using regular expressions to parse out the sequence. To implement the retrieval of sequences from GenBank by GI numbers we create a package called `org.jfb.util.GenbankDB`.

The `GenbankDB` class implements a method called `getSequence()` to retrieve sequences from GenBank through requests sent to the following URL (as defined in the String constant GENBANK_URL):

```
"http://www.ncbi.nlm.nih.gov/entrez/viewer.fcgi?dopt=fasta&
list_uids=";
```

Since the GenBank Id is a number is a numeral, the method performs checks if the user entered GI number is a valid entry. The `getSequence()` method takes a single parameter – the GenBank ID – opens a connection to the URL, obtains the data from GenBank, and performs the necessary parsing, formatting and trimming to get the actual GenBank sequence. To retrieve the CFTR sequence from GenBank using its GI number (6995995, replaced by 90421312), for example, the URL we would use in a browser would be: <<<here

```
http://www.ncbi.nlm.nih.gov/entrez/viewer.fcgi?dopt=fasta&l
ist_uids=6995995
```

This opens up the GenBank page with the sequence displayed in Fasta format (**Fig. 3.17**). This record needs to be parsed to extract the raw sequence from the HTML formatting on the page. This is easily done since the sequence is bounded by the <pre> and </pre> starting and ending tags. **Fig. 3.18** shows the source HTML of the page with beginning <pre> tag just before the Fasta formatted sequence starts.

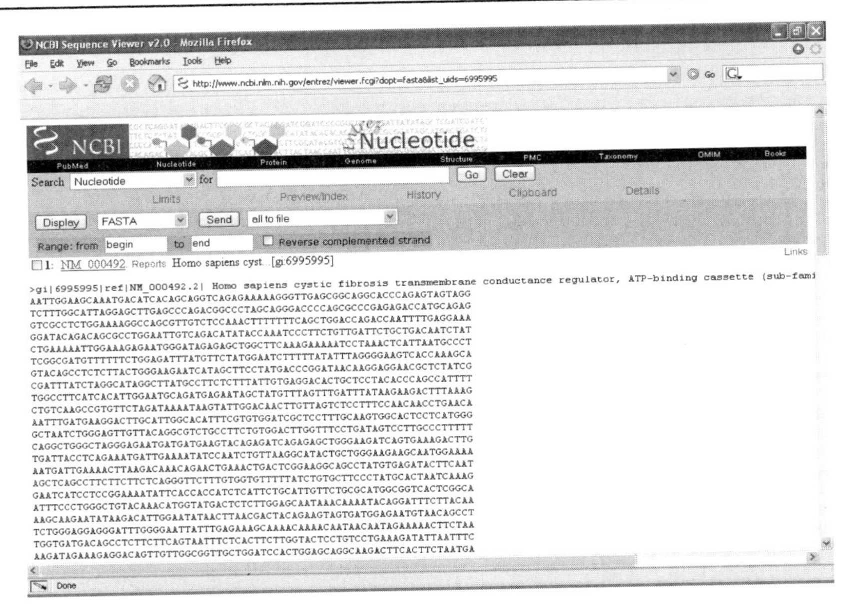

Fig. 3.17. GenBank record for the CFTR mRNA sequence

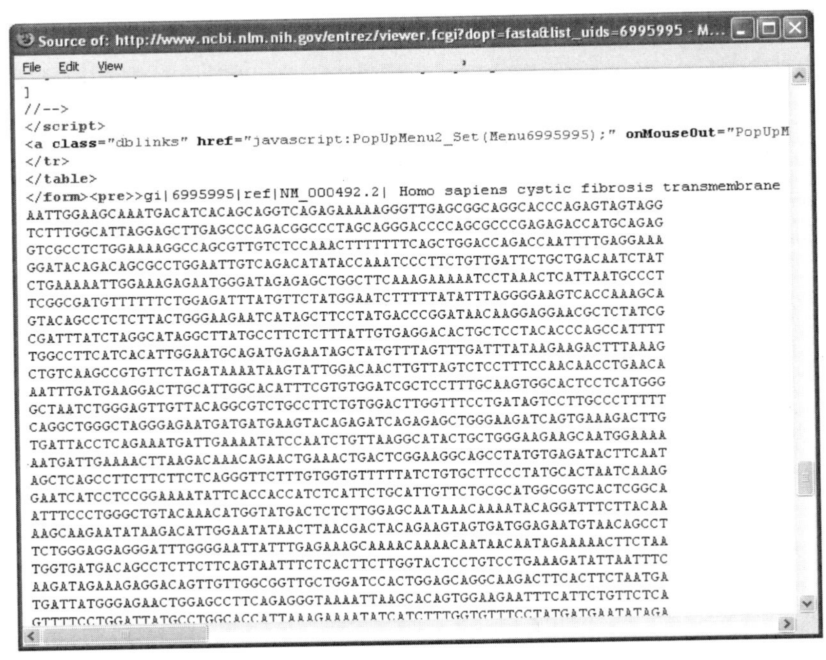

Fig. 3.18. Parsing the raw sequence data from a GenBank record

The code for the `GenbankDB` class is described in **Listing 3.9.**

Listing 3.9. The GenbankDB class

```
package org.jfb.util;
import org.apache.regexp.RE;
import org.apache.regexp.RESyntaxException;
import java.io.BufferedReader;
import java.io.IOException;
import java.io.InputStreamReader;
import java.net.URL;

public class GenbankDB {
    private static final String GENBANK_URL =
"http://www.ncbi.nlm.nih.gov/entrez/viewer.fcgi?dopt=fasta&li
st_uids=";

    public static String getSequence(String gbId) throws
IOException, IllegalArgumentException {
        // A GenBank ID is always a number
        boolean isGenBankID = isGenBankId(gbId);
        String genBankId = gbId.replaceAll("\n", "");

        if (!isGenBankID)
            throw new IllegalArgumentException(genBankId +
" is not a valid GenBank ID");

        BufferedReader reader = null;
        StringBuffer sb;

        try {
            URL url = new URL(GENBANK_URL + genBankId);
            reader = new BufferedReader(new
InputStreamReader(url.openConnection().getInputStream()));
            String s;
            sb = new StringBuffer();
            while ((s = reader.readLine()) != null) {
                sb.append(s + "\n");
            }
        } finally {
            if (reader != null)
                reader.close();
        }

        String tmp = sb.toString().toLowerCase();
        int idx = tmp.indexOf("<pre>");
        int endIdx = tmp.indexOf("</pre>");

        if (idx == -1 || endIdx == -1)
            return null;
```

```
            return    sb.substring(idx    +    "<pre>".length(),
endIdx);
      }

      private static final int CUT_OFF = 30;

      public static boolean isGenBankId(String gbId) {
          RE re = null;
          try {
              re = new RE("([0-9])+");
          } catch (RESyntaxException e1) {

          }

          boolean valid = true;
          String cleanSeq = gbId.replaceAll("\n", "");
          int len = cleanSeq.length();
          final int min = Math.min(CUT_OFF, len);
          String seqPiece = cleanSeq.substring(0, min);

              re.match(seqPiece);
              String match = re.getParen(0);
              valid     =     match     !=     null     &&
match.equals(seqPiece);
          return valid && min == len;
      }
  }
```

Input Validation

Note that there is no input validation in SwingBlast 2.2. SwingBlast 2.2 does not flag an error when bad characters are present in the sequence entered in the text area. **Fig. 3.19** shows the application behavior when an amino acid ("D") is inserted in what is apparently a nucleotide sequence. The sequence type is deduced as "N/A" because the application cannot determine the sequence type (**Listing 3.7**). For the same reason, none of the BLAST options are available. With the appropriate input validation, the application can catch errors in the entered sequence type and warn the user to make the appropriate changes.

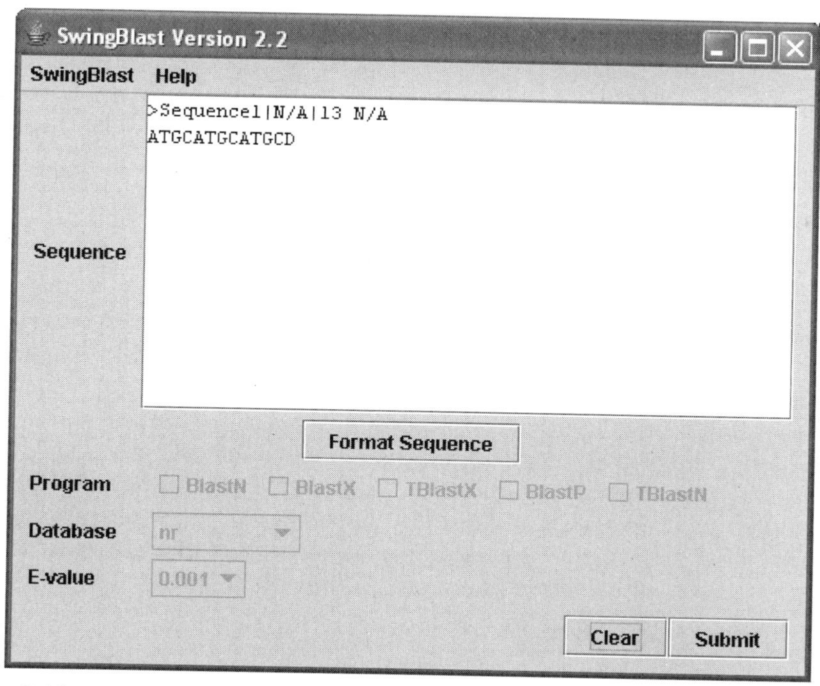

Fig. 3.19. Handling bad characters in input sequence

We will incorporate input validation for a few simple situations as described below:

1. The sequence contains bad characters, that is, characters other than the single letter codes for nucleotides and amino acids. We had illustrated how we used information on the composition of sequences found in nature to determine sequence type in Chapter 2. According to this algorithm, if:

   a. Total number of A, T, G and C's divided by the total length of the sequence is greater that 0.85, it is a DNA sequence

   b. Total number of A, T, G, C and U's divided by the total length of the sequence is greater that 0.85, it is an RNA sequence

If neither of these two conditions are met, the sequence is assumed to be a protein sequence. Again, we are not using the extended DNA/RNA alphabet that includes symbols for sequence ambiguity as defined in the

IUPAC-IUB nucleotide and amino acid nomenclature. Instead, we are illustrating input validation for the simplest of cases where the DNA alphabet is assumed to be composed of A, T, G, C and N and RNA is assumed to be A, U, G, C, N (where N = any nucleotide base) and amino acid alphabet is assumed to be A, C, D, E, F, G, H, I, K, L, M, N, P, Q, R, S, T, V, W and Y.

Once we have determined the sequence to be DNA, RNA or protein, we check if any bad characters are present in the sequence and warn the user accordingly. We also check if a number instead of a sequence has been entered in the text area. This may very well be a GI number. If it is indeed a GI number, the application will download the corresponding sequence from GenBank when the user presses the "`Format Sequence`" button. If none of the above conditions are met, the application will print an error message asking the user to check the validity of the sequence or data entered. To add input validation to the SwingBlast application, we add a method called `isValidSequence()`. The method takes the input sequence as the parameter and performs the appropriate checks as described earlier using regular expressions:

```
    private static boolean isValidSequence(String seq) {
      int idx = seq.indexOf(">");
      int idxEndOfFastaHeader = seq.indexOf("\n");
      String sequenceToCheck = null;
      if (idx != -1) {
        sequenceToCheck = seq.substring(idxEndOfFastaHeader +
1, seq.length());
      } else {
        sequenceToCheck = seq;
      }
      return matchRegex(REGEX_DNA, sequenceToCheck)
        || matchRegex(REGEX_RNA, sequenceToCheck)
        || matchRegex(REGEX_PROTEIN, sequenceToCheck)
        || matchRegex(REGEX_GENBANK_ID, sequenceToCheck);
    }
```

The regular expression matching within `matchRegex()` method checks for the following valid patterns:

```
    private static final String REGEX_DNA = "[acgtnACTGN]+";
    private static final String REGEX_RNA = "[acgunACUGN]+";
    private    static    final    String    REGEX_PROTEIN    =
"[acdefghiklmnpqrstvwyACDEFGHIKLMNPQRSTVWY]+";
    private static final String REGEX_GENBANK_ID = "[0-9]+";
```

The `matchRegex()` method itself is as follows:

```
    private static boolean matchRegex(String regex, String
sequence) {
      RE re = null;
      try {
        re = new RE(regex);
      } catch (RESyntaxException res) {
        // The regex has been tested so no need to chech the
        // exception here
      }

      String cleanSeq = sequence.replaceAll("\n", "");
      boolean valid = true;
      int len = cleanSeq.length();
      int pvsIdx = 0, nextIdx;
      for (int i = 0; i < len; i += CUT_OFF) {
        nextIdx = Math.min(i + CUT_OFF, len);
        String    seqPiece    =    cleanSeq.substring(pvsIdx,
nextIdx);
        re.match(seqPiece);
        String match = re.getParen(0);
        valid = match != null && match.equals(seqPiece);

        if (!valid)
          break;
        pvsIdx = nextIdx;
      }
      return valid;
    }
```

Next we call the `isValidSequence()` method in the `actionPerformed()` event method:

```
        public void actionPerformed(ActionEvent e) {
          // Check sequence type
          // Retrieve text entered in the text area
          final String sequenceText = sequenceArea.getText();
          if (sequenceText == null || sequenceText.length()
== 0) {
            cleanAllParameters();
            return;
          }

          if (!isValidSequence(sequenceText)) {
            JOptionPane.showMessageDialog(SwingBlast2_2.this,
              "The sequence you've entered is neither a DNA
or protein sequence nor a FASTA formatted sequence.\n" +
              "Please provide a valid sequence.");
            return;
          }
```

The application is now able to detect errors in the entered sequence and warn the user with the appropriate message (**Fig. 3.20**).

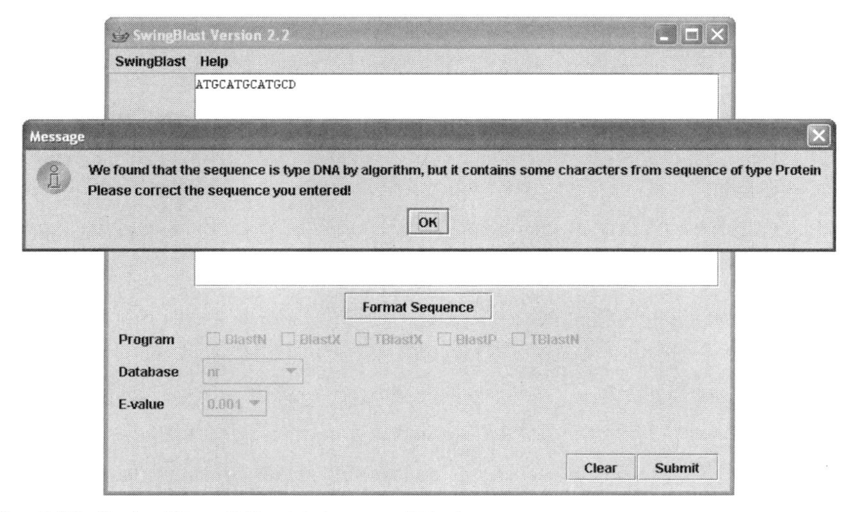

Fig. 3.20. SwingBlast 2.2 with input validation

Fig. 3.21 shows that the application recognizes that just a Fasta header has been provided and results in an error.

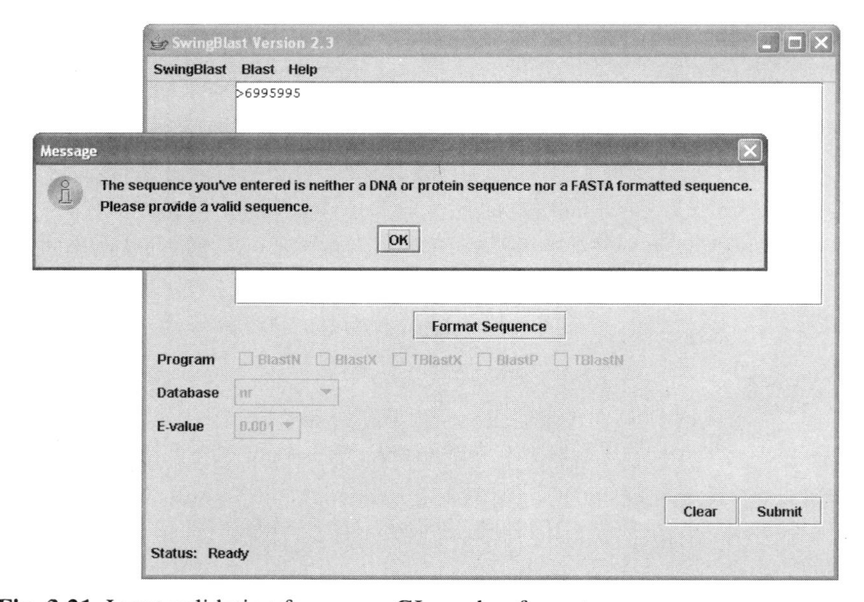

Fig. 3.21. Input validation for wrong GI number format

The application, however, does retrieve the correct sequence information from GenBank if a GI number is provided.

Controlling Program Events and Responses

We will next incorporate some program flow features in `SwingBlast` 2.2. We will program the "`Format Sequence`" button to be enabled only when a non-Fasta formatted sequence is entered in the text area. The format button will be disabled when the application starts and also under the following conditions:

1. When no sequence is available in the text area
2. If the sequence is already in Fasta format (at the time of pasting or right after the sequence is Fasta formatted)
3. When the "`Clear`" button is pressed
4. When the "`Format Sequence`" button is pressed

Let's enhance our application with these features in mind. We will call this `SwingBlast` Version 2.3. The code to enable or disable the "`Format Sequence`" button to meet condition #1 stated above is straight forward as shown in the **Listing 3.10** below. We implement a document listener interface and associate it with the text area widget. Within the document listener, we implement the `insertUpdate()` and `removeUpdate()` methods to respond to events that either insert or modify text within the text area. **Fig. 3.22** shows the `SwingBlast` 2.3 application with the format button disabled at launch.

Listing 3.10. Enabling and disabling the Format button

```
private void addListeners() {
  docListener = new DocumentListener() {
    public void insertUpdate(DocumentEvent e) {
      String text = sequenceArea.getText();
      if (text == null || text.length() == 0) {
        enableFunctions(-1);
        formatBtn.setEnabled(false);
      } else
        formatBtn.setEnabled(true);
    }
```

```
        public void removeUpdate(DocumentEvent e) {
          String text = sequenceArea.getText();
          if (text == null || text.length() == 0) {
            enableFunctions(-1);
            formatBtn.setEnabled(false);
          }
        }

        public void changedUpdate(DocumentEvent e) {
        }
      };

sequenceArea.getDocument().addDocumentListener(docListener);
```

Similarly, to meet condition 2, we include the following code:

```
  final boolean fastaFormatted = sequenceText.indexOf(">") !=
-1;
  formatBtn.setEnabled(!fastaFormatted);
```

For condition 3, the code is as follows:

```
  clearBtn.addActionListener(new ActionListener() {
  public void actionPerformed(ActionEvent e) {
      cleanAllParameters();
    }
  });
```

The cleanAllParameters() method will empty the sequenceArea and by doing this the document listener shown ealier will disable the formatButton as shown below:

```
  private void cleanAllParameters() {
       sequenceArea.setText("");
       enableFunctions(-1);
  }
```

Finally, to meet condition 4, when the button is pressed, the action listener is actually disabling the format button:

```
  formatBtn.addActionListener(new ActionListener() {
       public void actionPerformed(ActionEvent e) {
              ...

              formatBtn.setEnabled(false);
       }
```

```
});
```

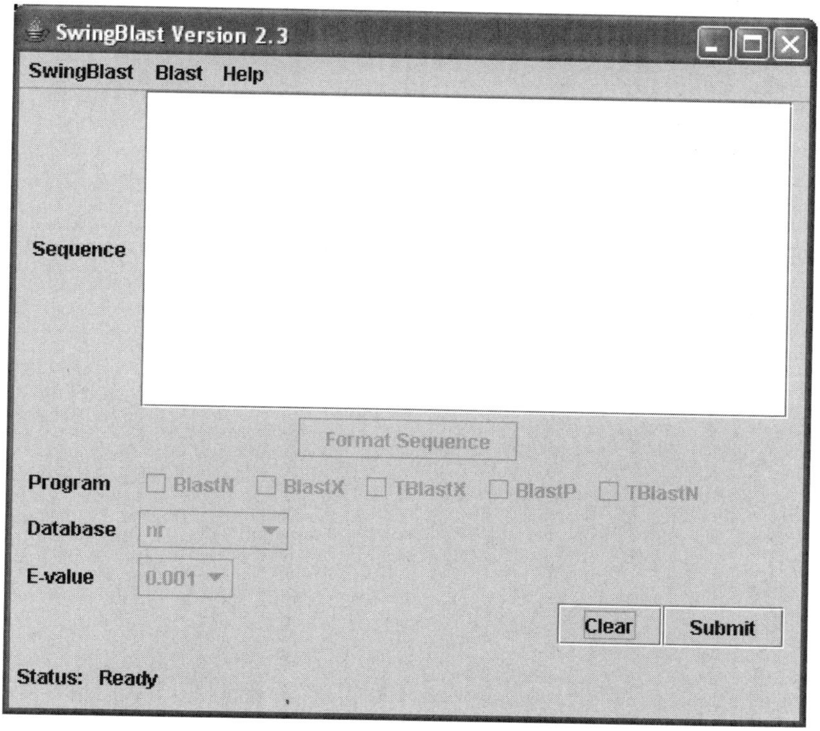

Fig. 3.22. Format button is disabled at start-up

Reporting BLAST Status

SwingBlast allows users to send sequences for multiple simultaneously BLAST analyses. It would be very informative to the user if the application were to provide a status of the current job that it is performing. In the next version of the application, we will add a program status bar to do so. With the program status code in place, the application will provide the user a running status of the jobs in process. Note that these messages will be relayed directly from the QBlast service and printed on the status bar using the observable method discussed earlier. The `SwingBlast` 2.3 application starts with the "`Status: Ready`" message at the bottom left of the application window as shown in **Fig. 3.23**. **Fig. 3.24** shows the status while the application is retrieving a sequence from GenBank based on a GI number. **Fig. 3.25** and **Fig. 3.26** show the status immediately after submitting a BLAST search and an intermediate stage before getting the results back. After all searches are complete, the system returns to the "`Ready`" status.

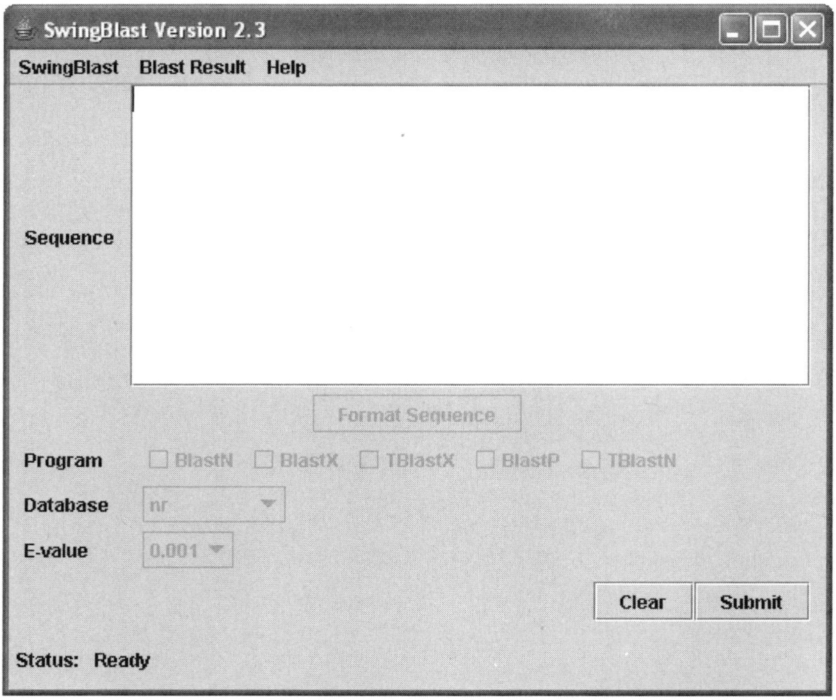

Fig. 3.23. Printing BLAST search status at start-up

Fig. 3.24. SwingBlast status during sequence retrieval from GenBank

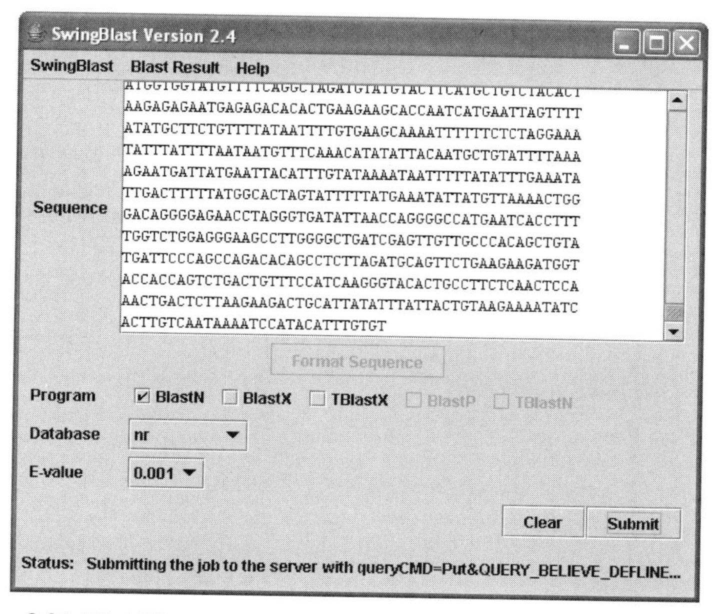

Fig. 3.25. BLAST search status at the time of submission

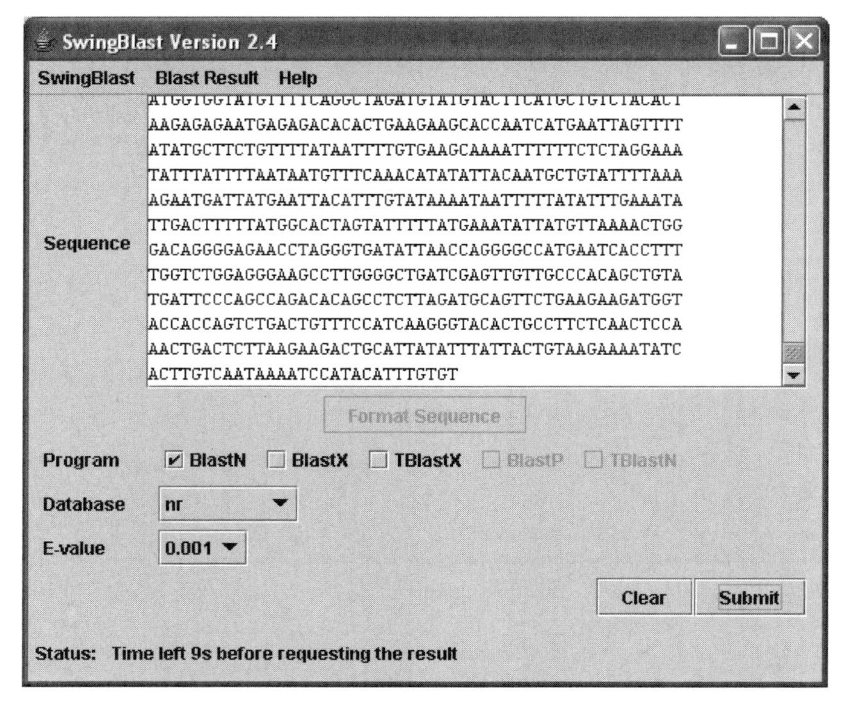

Fig. 3.26. Printing BLAST search status before getting results

The code for adding BLAST search status is as follows. First, we add a status label at the bottom left side of the application:

```
        statusLabel = new JLabel(STATUS_LABEL);
        statusLabel.setPreferredSize(new Dimension(50, 30));
        statusText = new JLabel(STATUS_READY);
        JPanel statusPanel = new JPanel();

statusPanel.setBorder(BorderFactory.createEmptyBorder(0,    5,
5, 5));
        statusPanel.setLayout(new BorderLayout());
        statusPanel.add(statusLabel, BorderLayout.WEST);
        statusPanel.add(statusText, BorderLayout.CENTER);


   newContentPane.add(statusPanel, BorderLayout.SOUTH);
```

If a GI number has been entered in the sequence area, the application gets the corresponding sequence from GenBank and displays the appropriate status message as shown in **Fig. 3.24**. The code for implementing this is shown below:

```
if (isGenBankID) {
     boolean canGetSeq = true;
     final String statusText = "Retrieving sequence for
GenBank ID " + sequenceText;
     try {
          SwingUtilities.invokeAndWait(new Runnable() {
               public void run() {

SwingBlast2_3.this.statusText.setText(statusText);
               }
          });
          seqObject = genbankSequenceDB.getSequence(text);
     } catch (Exception e) {
          e.printStackTrace();
     }
     sequence = seqObject.seqString();
          SwingUtilities.invokeAndWait(new Runnable() {
               public void run() {
                    resetStatusText();
               }
          });
     } catch (IllegalArgumentException iae) {
          // ignore because we checked already!
     } catch (Exception e) {
     canGetSeq = false;
}
if (seq == null || seq.length() == 0 || !canGetSeq) {
     JOptionPane.showMessageDialog(SwingBlast2_3.this,
          "Cannot get the sequence for GenBank ID " +
sequenceText);
     return;
}
```

Displaying BLAST Results Interactively

Finally, we will enhance the display capabilities of `SwingBlast` so we can view the results of the BLAST in a graphical and interactive manner. We will call this `SwingBlast` Version 2.4. The application will appear as shown in **Fig. 3.27**.

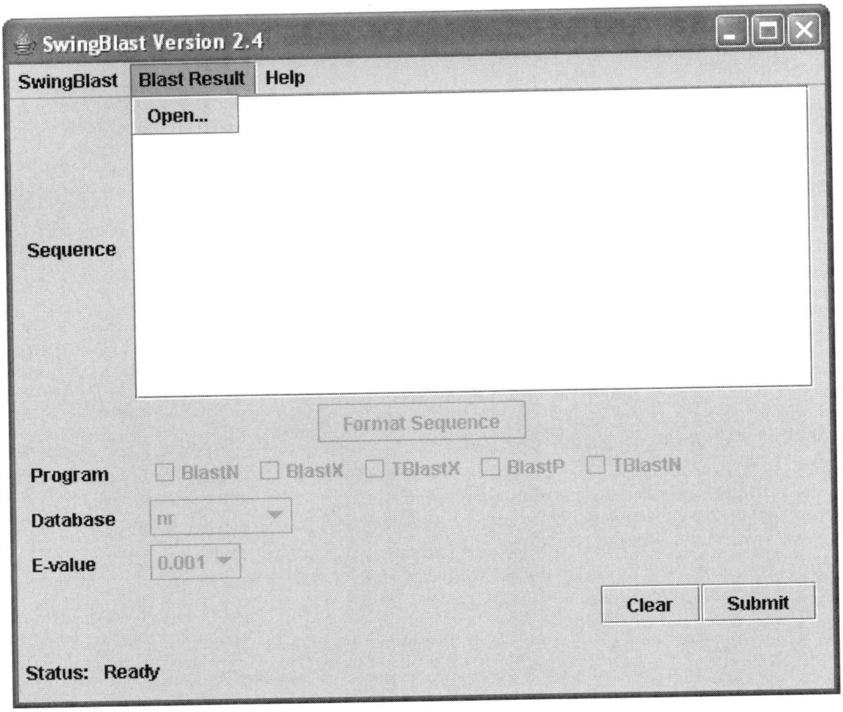

Fig. 3.27. Displaying BLAST results interactively

As seen in the **Fig. 3.27**, the user will select the `Blast` → `Open` menu button to access saved BLAST results. This will open a new window that will display the results in an interactive format. The data displayed in the graphical view is obtained by parsing information from the XML output shown in **Fig. 3.28.** The data parsed from the XML file includes such fields as the Hit_id, Hit_definition, Hit_accession, Hit_len, Hit_hsps, Hsp_number, Hsp_bit-score, Hsp_score, Hsp_evalue, etc. These fields describe the various attributes of a *High Scoring Sequence Pair* (HSP), such as the id, the definition, the GenBank accession number, the length, score, E-value etc. An HSP is a pair of aligned sequences of arbitrary but equal length, one derived from the query (input) sequence and one derived from the database it was searched against, that was returned by the BLAST search. The HSPs represent sequences whose alignment is locally maximal and for which the alignment score meets or exceeds a threshold or cutoff score provide by the user.

Fig. 3.28. XML file containing BLAST results data

BLAST results displayed in an interactive format are shown in **Fig. 3.29.**

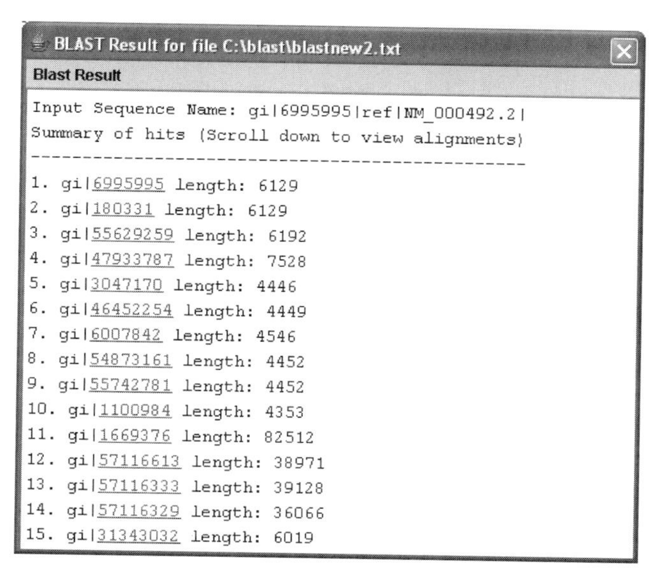

Fig. 3.29. Displaying BLAST results in an interactive format

Clicking on the GI number opens the GenBank record (**Fig. 3.30**).

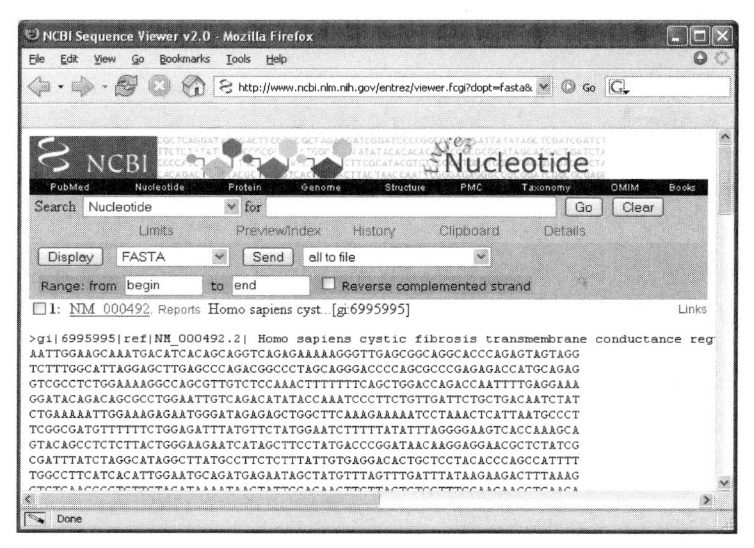

Fig. 3.30. Accessing GenBank record from BLAST results

Let's add the code that displays the BLAST results interactively. First, we add a menu item "BLAST Result" in the menu bar:

```
JMenu blastMenu = new JMenu("Blast Result");
openItem = new JMenuItem("Open...");
blastMenu.add(openItem);
menu.add(blastMenu);
```

Next we create a method called `displayBlastResult()` that takes a file name containing the BLAST results (that we saved earlier) as a parameter. The code is shown below:

```
      private     void     displayBlastResult(final     String
blastFileName) {
      final JDialog blastDialog = new JDialog(this, "BLAST
Result for file " + blastFileName, false);
      final JTextArea textArea = new JTextArea();
      final Font sf = textArea.getFont();
      Font   f   =   new   Font("Monospaced",   sf.getStyle(),
sf.getSize());
      textArea.setFont(f);

      Runnable runnable = new Runnable() {
        public void run() {
```

```
            Collection                  blastHits                  =
extractBlastHits(blastFileName);
            final String text = createReport(blastHits, new
ColorFormatterDNA());
            SwingUtilities.invokeLater(new Runnable() {
              public void run() {
                textArea.setText(text);
              }
            });
          }
        };
        new Thread(runnable).start();
        textArea.setLineWrap(true);
        final JMenuBar menuBar = new JMenuBar();
        JMenu menu = new JMenu("Blast Result");
        blastDialog.setJMenuBar(menuBar);
        JMenuItem openItem = new JMenuItem("Open...");
        openItem.addActionListener(new ActionListener() {
          public void actionPerformed(ActionEvent e) {
            final String blastResult = getBlastFileFromUser();
            if (blastResult != null)
              displayBlastResult(blastResult);
          }
        });
        menu.add(openItem);
        JMenuItem menuItem = new JMenuItem("Close");
        menuItem.addActionListener(new ActionListener() {
          public void actionPerformed(ActionEvent e) {
            closeMenu(blastDialog);
          }
        });
        menu.add(menuItem);
        menuBar.add(menu);
        blastDialog.addWindowListener(new WindowAdapter() {
          public void windowClosing(WindowEvent e) {
            closeMenu(blastDialog);
          }
        });
        blastDialog.getContentPane().add(new
JScrollPane(textArea));
        blastDialog.setSize(CP_PREF_SIZE);
        centerLocation(blastDialog);
        blastDialog.setVisible(true);
    }
```

Next we add a method to create the report called `createReport()` which takes the *Collection* object and a color formatter object called `ColorFormatter`:

```
      private   String   createReport(Collection   blastHits,
ColorFormatter colorFormatter) {
            StringBuffer           summary           =           new
```

```
StringBuffer("<html><body style=\"font-family: 'Monospaced',
Courier\">" +
                  "Input Sequence Name: " + inputSeqName +
"\n");
          StringBuffer alignments = null;
          if (blastHits == null || blastHits.size() == 0) {
              summary.append("No hits found from BLAST");
          } else {
              summary.append("Summary of hits (Scroll down to
view alignments)\n" +
                      "-----------------------------------------
----------\n");
              BlastHit hit;
              BlastHsp hsp;
              Iterator iterator = blastHits.iterator();
              alignments = new StringBuffer("\nAlignments\n"
+
                      "----------\n");
              int i = 1;
              while (iterator.hasNext()) {
                  hit = (BlastHit) iterator.next();
                  String hitId = hit.getHitId();
                  String genbankId = getGenBankId(hitId);
                  StringBuffer tmp = new StringBuffer("" +
i++)
                          .append(".                    gi|<a
href=\"http://www.ncbi.nlm.nih.gov/entrez/viewer.fcgi?dopt=fa
sta&list_uids="         +          genbankId            +
"\">").append(genbankId).append("</a>               length:
").append(hit.getHitLen())
                          .append("\n");
                  summary.append(tmp);
                  alignments.append(tmp);
                  Iterator hspIte = hit.getHsps().iterator();
                  while (hspIte.hasNext()) {
                      hsp = (BlastHsp) hspIte.next();
                      alignments.append("Score                =
").append(hsp.getBitScore()).append(" bits  E-Value: ")

.append(hsp.getEvalue()).append("\n\n");
                      int             queryFrom              =
Integer.parseInt(hsp.getQueryStart());
                      int              queryTo               =
Integer.parseInt(hsp.getQueryEnd());
                      int            subjectFrom             =
Integer.parseInt(hsp.getSubjectStart());
                      int              subjecTo              =
Integer.parseInt(hsp.getSubjectEnd());
                      appendSequences(alignments,
hsp.getQseq(),      hsp.getMidline(),       hsp.getHseq(),
NUMB_OF_CHAR_PER_LINE,
                              queryFrom,            queryTo,
subjectFrom,  subjecTo,  queryFrom  <  queryTo,  subjectFrom  <
```

```
subjecTo, colorFormatter);
                        alignments.append("\n");
                    }
                    alignments.append("\n");
                }
                summary.append(alignments);
            }
            return
summary.append("</body></html>").toString().replaceAll("\n",
"<BR>");
        }
```

The color formatter that adds colors to the DNA alignment is as follows:

```
    private class ColorFormatter implements ColorFormatter {
       public String format(String s) {
          String upperCaseSeq = s.toUpperCase();
          String color;
          String letter;
          for (int i = 0; i < letters.length; i++) {
            letter = letters[i];
            color = letterToColor.get(letter);
            upperCaseSeq  =  upperCaseSeq.replaceAll(letter,
"<font color=\"" + color + "\">" + letter + "</font>");
          }
          return upperCaseSeq;
       }
    }
```

The output of the program, which we call SwingBlast 2.5, is shown in **Fig. 3.31-3.32**. **Fig. 3.1** shows a high scoring alignment with a top hit (no gaps) and **Fig. 3.32** shows alignment with a sequence with a lower score (with gaps).

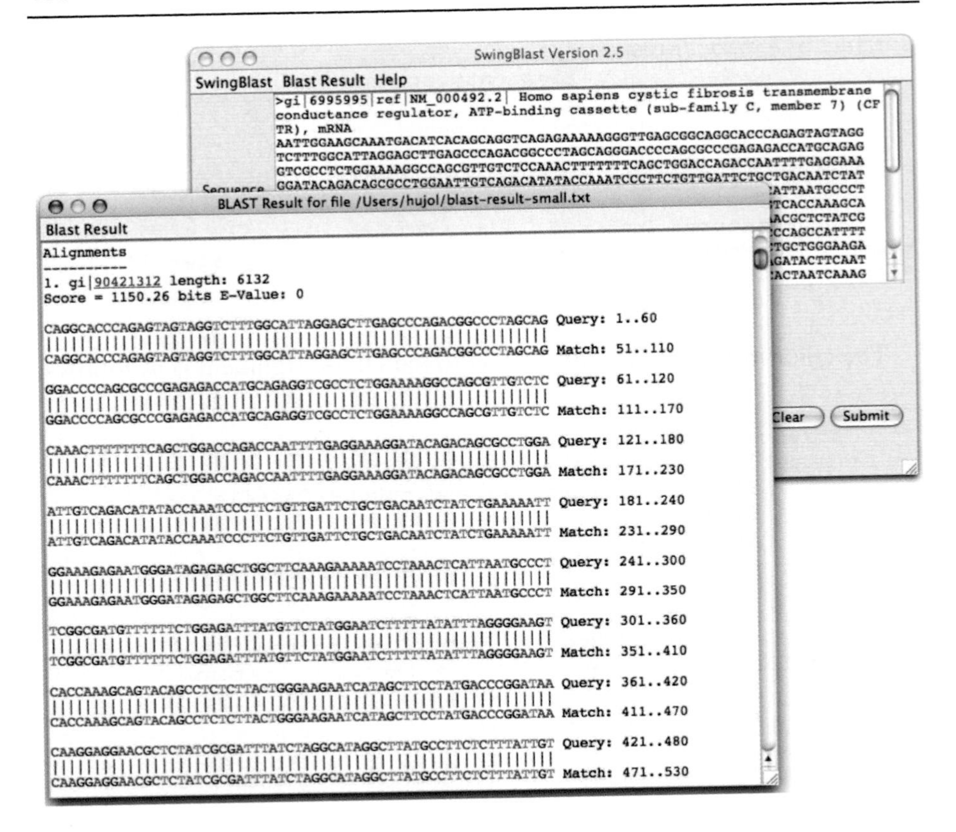

Fig. 3.31. Alignment without gaps

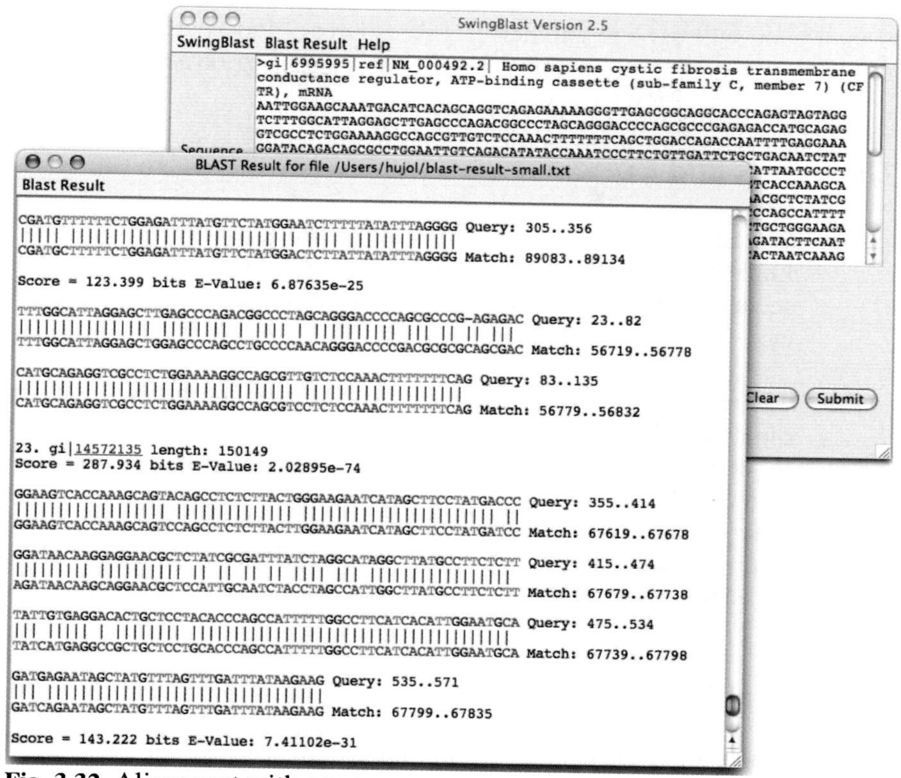

Fig. 3.32. Alignment with gaps

Summary

In this Chapter we have demonstrated the development of a complete BLAST application using the NCBI QBlast package. We created BLAST API and demonstrated how they could be used for BLAST analysis using a user interface, which allows users to send sequences to the QBlast service. We demonstrated the use of existing BioJava libraries to retrieve sequences from GenBank. We also enhanced the BLAST search output by allowing users to link returned hits to GenBank and to view alignments in color. The NCBI BLAST service is an indispensable resource for biomedical research and is frequently among the first analytic tool that is used in routine research investigations. The purpose of this Chapter was to provide the user with a comprehensive understanding of the resource as well as to demonstrate how J2EE can be used to develop user-friendly applications to simplify this fundamental research activity. In the next

Chapter, we will explore another useful resource – PubMed and expose a different aspect of Java – namely, JavaServer Pages and Java Servlets.

Questions and Exercises

1. We have built SwingBlast to retrieve sequences from GenBank. Enhance the application by including the functionality to retrieve sequences from other data sources such as Ensembl, Swiss-Prot, etc.

2. The aim of BLAST searches is to provide information on the biological function of an unknown piece of nucleotide or protein sequence. Write an application that takes the basic SwingBlast framework and provides information on the returned hits from other functional data sources such as Entrez Gene, UniGene, Gene Expression Omnibus (GEO), HomoloGene, OMIM (Online Mendelian Inheritance in Man™), etc.

Additional Resources

- Ensembl - http://www.ensembl.org/index.html
- Entrez Gene - http://www.ncbi.nlm.nih.gov/entrez/query.fcgi?db=gene
- GEO - http://www.ncbi.nlm.nih.gov/projects/geo/
- HomoloGene - http://www.ncbi.nlm.nih.gov/entrez/query.fcgi
- OMIM - http://www.ncbi.nlm.nih.gov/entrez/query.fcgi?db=OMIM
- Swiss-Prot - http://www.expasy.org/sprot/
- UniGene - http://www.ncbi.nlm.nih.gov/entrez/query.fcgi?db=unigene
- QBlast - http://www.ncbi.nlm.nih.gov/BLAST/Doc/urlapi.html

Selected Reading

UniGene: a unified view of the transcriptome. Pontius JU, Wagner L, Schuler GD. In: The NCBI Handbook. Bethesda (MD): National Center for Biotechnology Information; 2003.

NCBI GEO: mining millions of expression profiles - database and tools Tanya Barrett, Tugba O. Suzek, Dennis B. Troup, Stephen E. Wilhite, Wing-Chi Ngau, Pierre Ledoux, Dmitry Rudnev, Alex E. Lash, Wataru Fujibuchi and Ron Edgar. Nucleic Acids Research, 2005, Vol. 33, Database issue D562-D566.

An Overview of Ensembl. Ewan Birney, T. Daniel Andrews, Paul Bevan, Mario Caccamo, Yuan Chen, Laura Clarke, Guy Coates, James Cuff, Val Curwen, Tim Cutts, Thomas Down, Eduardo Eyras, Xose M. Fernandez-Suarez, Paul Gane, Brian Gibbins, James Gilbert, Martin Hammond, Hans-Rudolf Hotz, Vivek Iyer, Kerstin Jekosch, Andreas Kahari, Arek Kasprzyk, Damian Keefe, Stephen Keenan, Heikki Lehvaslaiho, Graham McVicker, Craig Melsopp, Patrick Meidl, Emmanuel Mongin, Roger Pettett, Simon Potter, Glenn Proctor, Mark Rae, Steve Searle, Guy Slater, Damian Smedley, James Smith, Will Spooner, Arne Stabenau, James Stalker, Roy Storey, Abel Ureta-Vidal, K. Cara Woodwark, Graham Cameron, Richard Durbin, Anthony Cox, Tim Hubbard, and Michele Clamp. Genome Res. 2004 May; 14(5):925-928.

Chapter IV

Facilitating PubMed Searches: JavaServer Pages and Java Servlets

Introduction

J2EE is a powerful platform for developing sophisticated web-based applications. This *J2EE* feature is especially critical for Bioinformatics software development given the availability of a large number of important biological sequence and biomedical data repositories on the WWW that biologists need to access on a routine basis for their research. We will explore one such resource - NCBI PubMed - in detail in this Chapter and introduce the *Java Servlet* and *JavaServer Pages (JSPs)* technologies to facilitate searching, retrieval and storage of biomedical data from PubMed.

HTTP and CGI

We will begin by refreshing our basic knowledge of standard protocols such as the Hypertext Transfer Protocol (HTTP) and the Common Gateway Interface (CGI) that allows for a server to pass requests from a client web browser to an external application and in return allow the web server to return the output from the application to the web browser. Although there are several more HTTP commands than GET and POST, we will introduce only these methods here and refer interested readers to the HTTP specification Request for Comments 2616 (RFC 2616) for more information.

HTTP Protocol

HTTP is a client/server protocol that WWW users utilize everyday to download web pages to their web browsers. The client part of this protocol is handled by the web browser that sends a request to the server (also called an HTTP server or a web server). The server responds to the request with a web page. That, put very simply is all that HTTP does, at least for the purpose of this discussion.

The request sent by the client contains an HTTP command with a set of parameters that define the request. For example, to request an HTML document called index.shtml from the NCBI server, one can issue the following command using telnet:

```
telnet www.ncbi.nlm.nih.gov 80
GET /blast/index.shtml HTTP/1.0
```

telnet is a program that connects a local computer to a server on the network and allows users to issue commands directly to the remote server. The HTTP protocol works over the Transmission Control Protocol/Internet Protocol, a suite of communications protocols used to connect hosts on the WWW, also called TCP/IP for short. In this case, the HTTP protocol works over the TCP/IP protocol that one can access through a session initiated by telnet, using the specified server address (www.ncbi.nlm.nih.gov) and the port (80).

There are other pieces of information that could be passed to the request, to specify information about the client and the type of data it would like to receive. Also a blank line specifying the end of the request must be added at the end.

When such a request is sent to the NCBI server, the output received contains several difference bits of data, along with the actual document requested, if found.

```
HTTP/1.1 200 OK
Date: Sun, 12 Feb 2006 18:13:42 GMT
Server: Nde
Accept-Ranges: bytes
Content-Type: text/html
Connection: close

<?xml version="1.0" encoding="UTF-8"?>
<!DOCTYPE    html    PUBLIC    "-//W3C//DTD    XHTML    1.0
```

```
Transitional//EN"
      "http://www.w3.org/TR/xhtml1/DTD/xhtml1-
transitional.dtd">
```

(The output has been truncated for clarity.)

The first line corresponds to a code indicating the status of the response - 200 OK - which means the requested operation was executed successfully. After the status line we have information about the server itself. Finally if the document is available it is sent within the rest of the response.Other code and associated descriptions are defined in the HTTP specification and provide information regarding any problems accessing the server, if the requested document is not found, etc.

GET and POST Methods

Although a client can send different HTTP commands, the GET and POST commands are the most commonly used. GET allows users to retrieve or get information from an HTTP server, while the POST HTTP command allows users to post or send information to the server. The POST information resides on the server, usually within a database. The GET command is just for querying the HTTP server and therefore won't be stored, unless for statistical purposes or for logging the load on the server.

GET can send parameters within the body of the URL to specifically query the HTTP server. Since GET was designed for querying purposes, the URL length is limited to a certain number of characters (250) on certain servers. The POST method, on the other hand, can send more information, including different documents types, and does not have a constraint on length.

CGI For Generating Dynamic Content

According to RFC 3875, CGI is a

"... simple interface for running external programs, software or gateways under an information server in a platform-independent manner."

This simply means that if you have a program that runs on your Unix machine and you want to access it through a web browser, you can do so using CGI. The way it works is that each time you request to run that program, the web server will create an instance of the program, pass to it all the parameters obtained from the request that was sent, wait for the program to process the information and then wrap the program output into an HTTP response.

This allows users to generate the content of a web page dynamically instead of accessing static HTML content. It can be very slow when 100 users access the same program because the server must create 100 instances of the same program to run the 100 queries.

A number of vendors have implemented their own API's to handle the performance issues of CGI or to replace that interface with proprietary protocols. Sun Microsystems, for example, has developed proprietary technology that will run in a Java Virtual Machine and handle the required processes that live on the server via the *Servlets* and *JavaServer Pages* technologies.

Servlets and JavaServer Pages Technologies

Now that we're more familiar with HTTP, it's time to learn about *servlets* and *JSPs*. Before we present the Java API, lets briefly review the advantages of using *servlets* over typical CGI programs:

- Once the *servlet container* is started, each *servlet* runs in the same process as the container; this avoids creating new processes for each request, unlike CGI programs.
- Because the *servlet* is created once at startup, it remains in memory and there is no overhead associated with loading the Java class multiple times. The service just needs to request the *servlet* from a pool and call its service method.
- A *servlet* is reusable, which saves memory and time.

These characteristics allow faster execution of the server processes to generate dynamic content. In addition, the fact that it is Java brings with it the power of the "Write once, run everywhere" properties of the platform.

Java API for Servlets and JSPs

From the *servlet* specification available at the Sun Microsystems website, a *servlet* is defined as a "Java technology-based Web component, managed by a container, that generates dynamic content". *Servlets* are Java classes that implement a base interface called `Servlet`, from the `javax.servlet` package available in the *Java Enterprise Edition Platform.* `javax.servlet.Servlet` is the basic interface which provides the `service()` method that handles a client request independently of the protocol used to communicate between the client and the server. To create a *servlet* one can directly implement this interface or extend `GenericServlet` or `HttpServlet`.

The life cycle of a servlet is managed through three methods:

- `init`: the container instantiates a `Servlet` object and calls init to initialize it.
- `service`: upon a client request, the container get the *servlet* and calls its `service` method.
- `destroy`: when the *servlet* is not in use any more, the container will call the `destroy` method.

Fig. 4.1 below shows the life cycle of a servlet (called MyServlet) when a client request comes to the container.

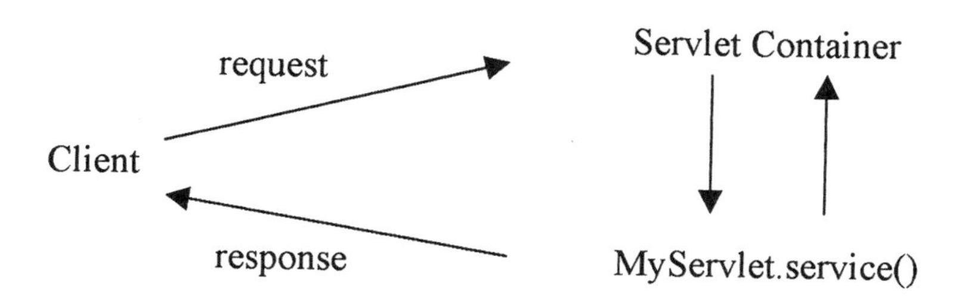

Fig. 4.1. Life cycle of servlets

Since we want to deal with HTTP requests, we are primarily interested in the `javax.servlet.http.HttpServlet` package to create `HttpServlet` Java classes. We will learn more about this package in the next few sections.

***

Before we delve into the servlet and JSP technologies, let's briefly review the MVC framework that we had introduced in Chapter 1, which we will be using as a guiding principle for building our web application. We will also briefly review the *Apache Tomcat Serve*r, which we will use as our *servlet container*. Finally we will also talk briefly about the *JavaServer Pages Standard Tag Library (JSTL),* to introduce the concept for the benefit of readers to explore further on their own.

JavaServer Pages Standard Tag Library (JSTL)

JavaServer Pages (JSPs) use *custom tags* to perform all kinds of manipulations like iterating over collections, transforming one object into another, form processing, database access, and the like. The idea behind *JSTL* is to create libraries with *reusable tags*. These tags can be used and customized like functions or methods in Java. This also creates clarity in the *JSP* file because the *tags* allow users to keep the *JSP* as the *View* and the *business logic* or the C*ontroller* and the *Model* separated from each other. In other words, one can think of *JSTL* as a Java package that groups together functionalities into a set of independent and reusable and tags.

Apache Tomcat Server

Tomcat is an open source *servlet container*, which implements the Java *Servlet* and *JavaServer Pages* technologies written in Java. This is the *servlet container* we will be using in this Chapter. The Tomcat servlet container allows developers to deploy web applications as well as to monitor and manage them. Tomcat compiles the *JSPs* into *servlets* when first called, or just before calling the application. Tomcat also allows defining the *realm* for specific authentication and authorization services that may be required for web applications. A "realm" in Apache terminology is "a "database" of usernames and passwords that identify valid users of a web application (or set of web applications), plus an enumeration of the list of roles associated with each valid user." The reader is referred to the Appendix for further information on how to install Tomcat. More information can also be found at the Apache Tomcat Project website of The Apache Software Foundation.

The NCBI PubMed Literature Search and Retrieval Service

PubMed is a resource maintained by the National Library of Medicine (NLM), under the aegis of the National Center for Biotechnology Information (NCBI, National Institutes of Health, USA) and provides access to over 14 million citations for biomedical articles dating back to the 1950's. PubMed is a vast resource and covers scientific findings from a diverse array of disciplines including but not limited to the natural and physical sciences. According to usage statistics from NCBI, over 59,000,000 queries seeking scientific information were submitted to the PubMed server in March 2004 alone (http://www.ncbi.nlm.nih.gov/About/tools/restable_stat_pubmed.html). Indeed, PubMed is an indispensable resource for researchers all over the world.

As vast and valuable as PubMed is, average users still have to contend with the problem of retrieving useful and relevant knowledge from the underlying database in a piecemeal fashion using one or more keywords. PubMed also doesn't currently provide a way to intelligently or visually analyze the results of a query (for example, by highlighting or color coding the search terms in a retrieved abstract, etc). We will address some of these issues and create solutions for them in this Chapter to enhance the value of literature search and retrieval through PubMed.

Accessing Biomedical Literature Through Entrez

Access to information in NCBI databases is granted through a service called Entrez, a search and retrieval system maintained by NCBI that combines information on individual DNA and protein sequences, large-scale sequence data from whole genomes, and information on 3-dimensional structures of biomolecules. It also grants access to MEDLINE, which covers research in a number of Life Science areas such as medicine, nursing, dentistry, veterinary medicine, the health care system, and preclinical sciences. The steps involved in a typical search on PubMed are described below. We will use the generic keyword "HIV" (for Human Immunodeficiency Virus, the causative agent of Acquired Immune Deficiency Syndrome, AIDS) for the illustration.

Step 1: User navigates to the NCBI PubMed website (**Fig. 4.2**):

http://www.ncbi.nlm.nih.gov/entrez/query.fcgi?db=PubMed

Step 2: User enters the search term 'HIV' (the search is case-insensitive) in the search box and presses Enter. PubMed presents the user with a list of citations relevant to the search term (**Fig. 4.3**). Internally, PubMed searches for a match between the supplied keyword(s) and terms in the Medical Subject Headings (MeSH) Translation Table, an alphabetical hierarchy of controlled vocabulary terms used for subject analysis of biomedical literature at the NLM. The list of citations may span several thousand pages depending on the number of articles that match the search term. Each journal article on PubMed is associated with a unique numeric tag called the PubMed Unique Identifier or PMID.

Step 3: User clicks on the citation to display specific information (Brief, Abstract, Medline etc) about each journal article (**Fig. 4.4**) or selects several articles to display (**Fig. 4.5**).

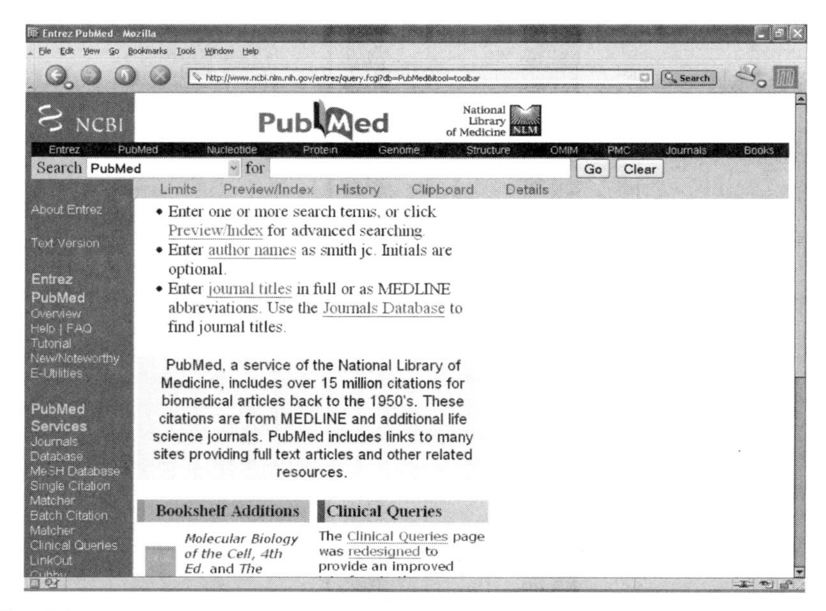

Fig. 4.2. The NCBI PubMed web resource

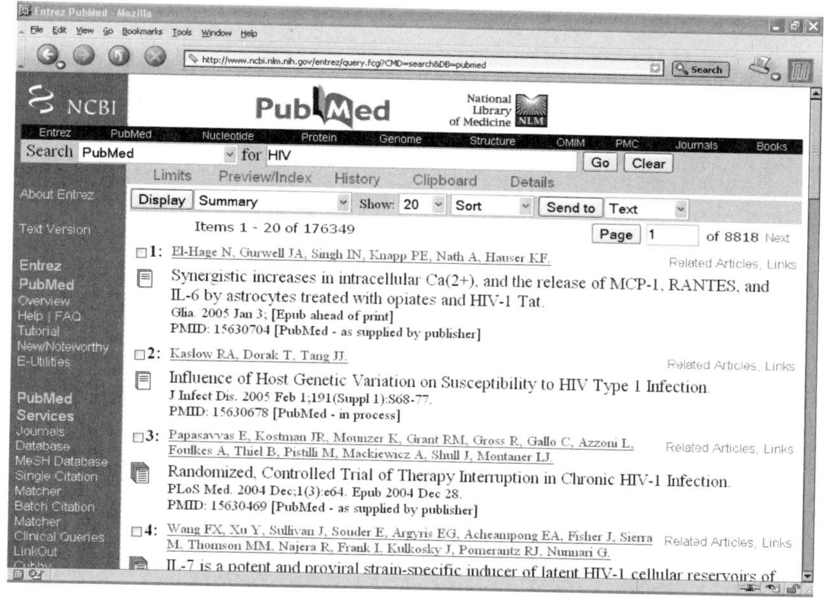

Fig. 4.3. Search results for the term 'HIV'

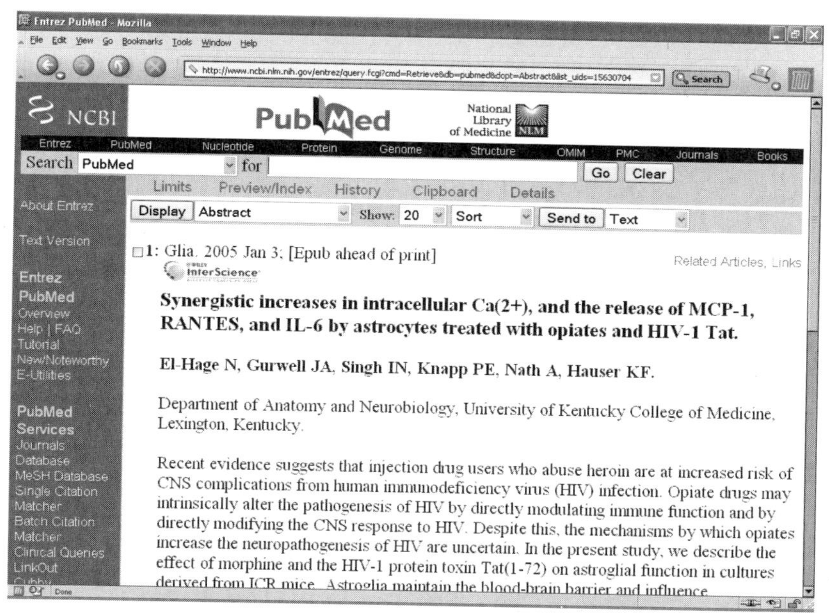

Fig. 4.4. Viewing abstracts for individual journal articles

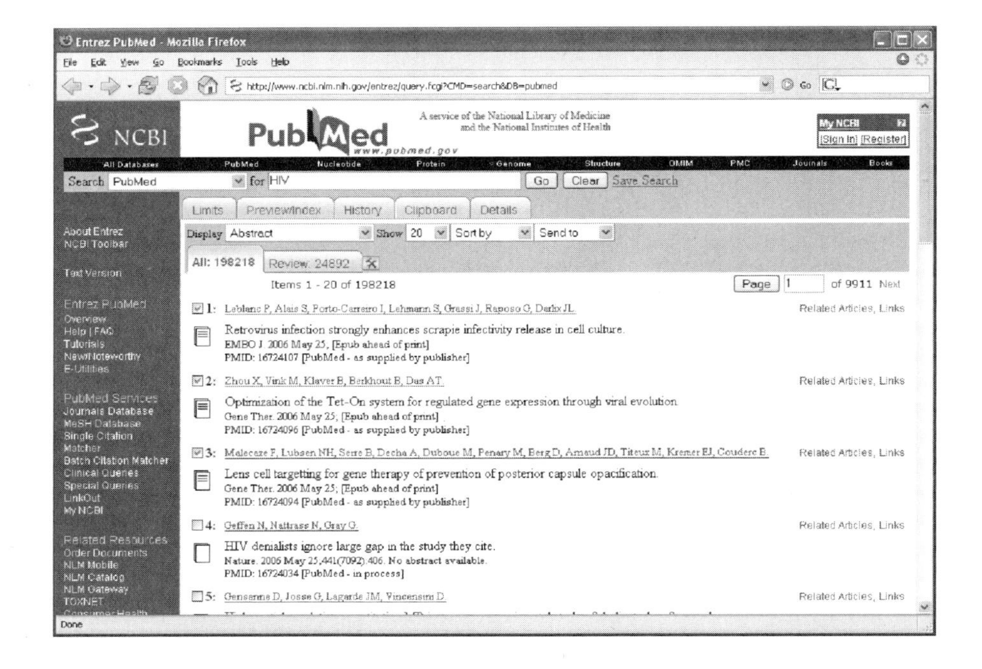

Fig. 4.5. Selecting several articles to view abstracts

The user can save articles of choice in the chosen display format (Summary, Abstract, etc) by selecting the required articles and pressing the "Send to" button and selecting the appropriate format (Text, File, Email, etc) (**Fig. 4.6**).

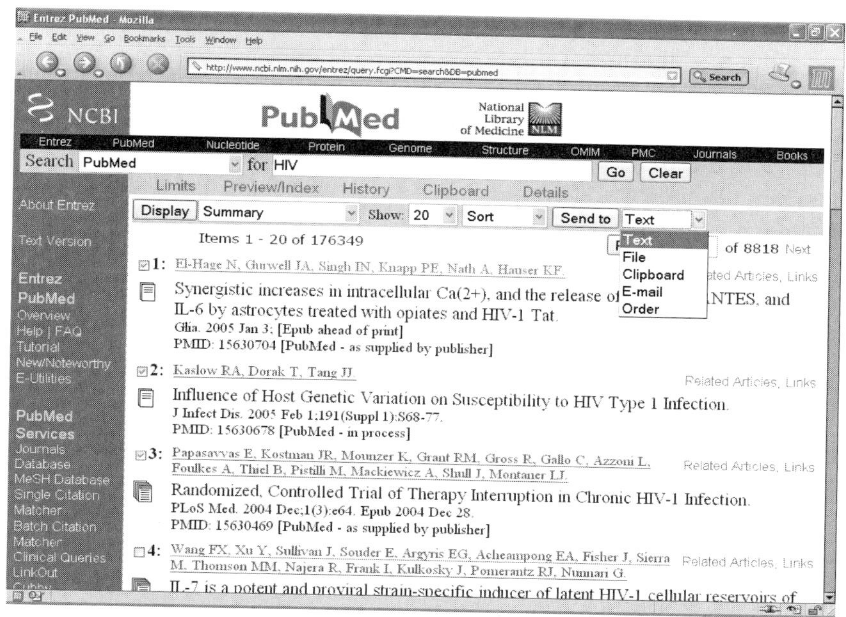

Fig. 4.6. Saving search results for selected abstracts

The search process quickly becomes unwieldy especially when information from a large number of citations needs to be extracted and analyzed. In this Chapter, we will demonstrate the power of Sun's JavaServer Pages and Java Servlets technologies to build a web-based application to simplify the process of accessing information on PubMed. We will use the Apache Tomcat server as the servlet container and the Apache Ant tool to build and deploy the Java web-based application. Please refer to the Appendix to download the tools and for instructions on using them.

Create Web Application With Servlets and JSPs

Servlets as we described earlier are Java code that run on a server and provide a general framework for services built using the request-response paradigm. HTTP, is one such paradigm that is implemented through the *javax.servlet.http* package from the Java Servlet API. On the other hand, *JSPs* were designed to mainly allow the separation of the business logic (what the application does) from the appearance of the page (how the application displays the result).

The steps and the flow diagram below illustrate the behavior of such an application (**Fig. 4.7**):

Step 1: The user accesses the application through a web browser. The actual code that runs the application remains hidden from view. The user only sees and interacts with an HTML page, which for our first application will contain a simple search form consisting of a single text-box and a submit button. The user enters a single keyword (search term) in the text-box and presses the submit button. After the search is processed by the application, the user sees the results in the web browser. **Fig. 4.7** illustrates the actions of the user in the User Space.

Step 2: The application is implemented as a *servlet* that gets the information entered on the search form and processes the request on the NCBI PubMed server. This involves a series of operations. The application constructs the PubMed URL that is specific to the entered search term. Next, through a URL object, it sends a request to the PubMed server. The PubMed server performs the search using the keyword and formulates a response, which is an HTML document containing a list of citations matching the search term. These operations are shown in the Application Space (**Fig. 4.7**).

Step 3: After processing the request, the PubMed server sends the search results back; the application reads the result from the URL using a `BufferedReader` object to retrieve the content sent back from the server.

Step 4: Once the response is received, the application reads the contents of the response using the `BufferedReader` object and prints it out to the screen using the `javax.servlet.http.HttpServletResponse` object.

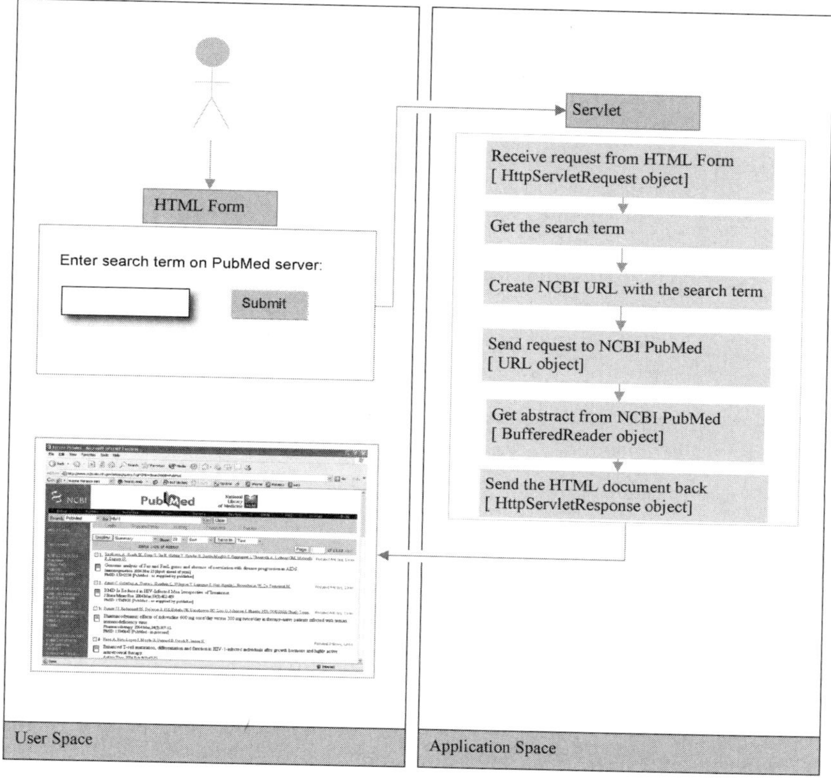

Fig. 4.7. The structure of the PubMed1 servlet

Web Application Structure

According to the Java Servlet API Specification 2.2 (http://java.sun.com/products/servlet/download.html), a web application (or web app) is a collection of *servlets*, HTML pages, classes, images, and other resources that can be bundled and run on multiple containers from multiple vendors. Simply stated, a web app bundles resources together to provide a portable and server independent way to access information via a web browser. In order to be portable and server independent, a web app must be designed according to a well-defined schema that dictates where the resources used by the web app are to be placed. This ensures that there is no conflict between the different resources used by the web app. The web app has to be installed on the web application server and mapped to a specific uniform resource identifier (URI) path (called also the *servlet context path*) on the server. The file structure of the web app is archived into a WAR file (Web application ARchive).

For example, the application we are writing is installed on the web app server using the path pubmed, for example:

```
http://localhost:8080/pubmed
```

as is explained further below. Here is the file structure of the `pubmed` web app being developed:

```
example.html
pubmedExample.jsp
jsp/moreSpecificPubmedExample.jsp
pics/pubmedLogo.png
anotherLogoExample.png
WEB-INF/web.xml
WEB-INF/classes/servlet/DataRetriever.class
WEB-INF/lib/Jakarta-regexp-1.3.jar
```

The basic layout that defines a web app file structure is as follows:

- HTML, JSP, PNG (image) and other resource files must be located in the root directory to be visible in the web browser.
- *web.xml* is located in the *WEB-INF* directory under root. *web.xml* is the Web Application Deployment Descriptor for the application. This file defines in an XML format the configuration information utilized by the web app such as initialization parameters, *servlet* mappings, security constraints, etc.
- *WEB-INF/classes*: This directory contains all the Java classes (and *servlets*) with any resources associated with them that make the web app. The Java class `servlet.DataRetriever` is stored in `WEB-INF/classes/servlet/DataRetriever.class`.
- *WEB-INF/lib*: This directory contains all the Java™ Archive (JAR) files required to run the web app, including third parties libraries such as Jakarta-regexp-1.3.jar for regular expression matching.

Access to the web app or any resource from the web application server available at the localhost and port 8080 is through the following URL:

```
http://localhost:8080/pubmed
```

This access is set up in the *http.conf* configuration file located in the Tomcat '*conf*' directory. Any web application is deployed on the web application server using a relative path.

If we want to access the HTML pages located in the WAR file in the root directory, for instance, for a file called `example.html`, we open the following URL in the web browser:

```
http://localhost:8080/pubmed/example.html
```

The WAR archive may also contain images that can be found in the `/pics` directory. To access the `pubmedLogo.png` picture, for example, we need to point our web browser to the following URL:

```
http://localhost:8080/pubmed/pics/pubmedLogo.png
```

To access the *servlet* `DataRetriever`, in the web app descriptor file we wrote the mapping from the path in the URL to the actual Java class that is going to handle the HTTP request. This Servlet can be accessed at the following path:

```
http://localhost:8080/pubmed/DataRetriever
```

Creating a Servlet to Access Biomedical Literature

We begin by declaring a package called `PubMed`. Next, we import the necessary packages, which contain the classes that are used by the *servlet*. In order to implement the design described in **Fig. 4.7**, we need to create a Java `Servlet` class called `PubMedServlet1_1` that extends `javax.servlet.http.HttpServlet`, the standard base class for HTTP *servlets*. We then need to override the `doGet()` method as shown in the code below. The `doGet()` method takes two parameters: the `HttpServletRequest` object (called req) which is the client request and an `HttpServletResponse` object (called res) which is the response sent back to the client. Since the method returns nothing, its return type is *void*.

It is conceivable that the process of sending a request to a remote server and obtaining a response back may encounter errors. Java has objects called *Exceptions* to handle such occurrences. The Java Virtual Machine (JVM) will inform the caller using *Exception* objects when a program does not behave the way it is supposed to do. This object is "*thrown*" when that error or unusual condition occurs and it stores information about the particular error event. In order to inform the developer that such an *exception* can be "thrown" from the method, we use the appropriately named "*throws*" Java keyword in the method signature. We declare `ServletException`, which defines a general *exception* a *servlet* can throw when it encounters errors and `IOException` to catch errors due to failed or interrupted I/O operations. Another way to handle *exceptions* is to use the *try-catch block*. We will see how to use *try-catch blocks* later in the Chapter.

Let's return to the servlet creation process. The general signature of the `doGet` method is shown below:

```
     Protected void
doGet(HttpServletRequest req, HttpServletResponse res)
            throws ServletException, IOException { }
```

since we are sending a text or HTML response, we set the content type to text/html with the line:

```
res.setContentType("text/html");
```

Next we request a `PrintWriter` object to write the text to the response message:

```
PrintWriter out = res.getWriter();
```

Next we create HTML to create a form that users can utilize for conducting searches on PubMed. In its simplest state, the form will have a title, a search box and a submit button. The HTML for the form is as follows:

```
<HTML>
  <HEAD><TITLE>PubMed Servlet 1.1</TITLE></HEAD>
  <BODY>
    <b>Java for Bioinformatics: </b>
      <font     color=red><b>PubMed     Servlet     version
1.1</b></font>\n
      <BR><BR><B>Please  enter  a  term  to  search  on  NCBI
PubMed:</B><BR><BR>\n
      <FORM METHOD=GET>\n
      <INPUT TYPE=TEXT NAME=searchTerm><BR><BR>\n
      <INPUT TYPE=SUBMIT VALUE=\"Search PubMed\"><BR>\n
    </FORM>
  </BODY>
</HTML>
```

The search form as it appears in a browser is shown in **Fig. 4.8**.

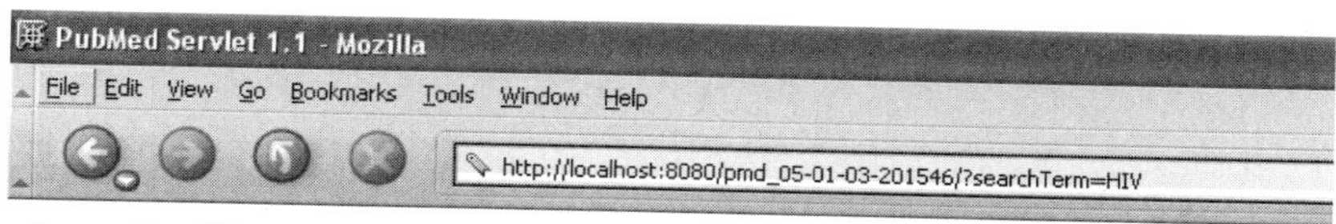

Fig. 4.8. The PubMed servlet version 1.1 search form

To implement the form in code, we create an object called html of the type `StringBuffer`:

```
StringBuffer html = new StringBuffer("<HTML>");
```

and append the HTML code to it:

```
StringBuffer html = new StringBuffer("<HTML>");
      html.append("<HEAD><TITLE>PubMed Servlet
1.1</TITLE></HEAD><BODY>\n");
      html.append("<b>Java for Bioinformatics: </b>");
      html.append("<font color=red><b>PubMed Servlet version
1.1</b></font>\n");
      html.append("<BR><BR><B>Please enter a term to search on NCBI
PubMed: </B><BR><BR>\n");
      html.append("<FORM METHOD=GET>\n");
      html.append("<INPUT TYPE=TEXT NAME=searchTerm><BR><BR>\n");
      html.append("<INPUT TYPE=SUBMIT VALUE=\"Search
PubMed\"><BR>\n");
      html.append("</FORM>\n");
```

The URL to send the search term is:

```
http://www.ncbi.nlm.nih.gov/entrez/query.fcgi?dispmax=10&db=pubmed&
cmd=search&term=term
```

In code we implement this in the following manner:

```
URL url = new URL
("http://www.ncbi.nlm.nih.gov/entrez/query.fcgi?dispmax=10&db
=pubmed&cmd=search&term=" + URLEncoder.encode(term, "UTF-
8"));
```

Note the parameters on the URL (separated by ampersand symbols '&') that specifies what information we want to submit to the PubMed engine to retrieve data:

```
dispmax=10
db=pubmed
cmd=search
term=term
```

We are limiting the search to ten articles (dispmax=10) for the purpose of illustration only. We select the database as PubMed (db=pubmed) and provide the command to search (cmd=search) with the search term (term=term). Next, we open the connection to the server:

```
URLConnection urlConnection = url.openConnection();
BufferedReader    reader    =    new    BufferedReader(new
InputStreamReader
     (urlConnection.getInputStream()));
```

In the next step, we construct a regular expression to extract the PubMed Ids (PMIDs) of the abstracts that match the search term and create an array to store them. To do this, we will use a Java Regular Expression package available from The Apache Jakarta Project available as a JAR file called jakarta-regexp-1.3.jar:

```
String s = null;
RE pmidRE = new RE("PMID: ([0-9]+) \\[PubMed");
Collection pmids = new ArrayList();

while ((s = reader.readLine()) != null) {
  if (pmidRE.match(s)) {
    pmids.add(pmidRE.getParen(1));
  }
}
reader.close();
```

Listing 4.1 shows the code for PubMed servlet version 1.1

Listing 4.1. PubMed Servlet version 1.1

```
package org.jfb.PubMed;
import org.apache.regexp.RE;
import javax.servlet.ServletException;
import javax.servlet.http.HttpServlet;
import javax.servlet.http.HttpServletRequest;
import javax.servlet.http.HttpServletResponse;
import java.io.*;
import java.net.URL;
import java.net.URLEncoder;
import java.net.URLConnection;
import java.util.ArrayList;
import java.util.Collection;
import java.util.Iterator;
import java.util.Properties;

public class PubMedServlet1_1 extends HttpServlet {
  protected void doGet(HttpServletRequest req,
HttpServletResponse res)
    throws ServletException, IOException {
    res.setContentType("text/html");
    PrintWriter out = res.getWriter();

    StringBuffer html = new StringBuffer("<HTML>");
    html.append("<HEAD><TITLE>PubMed Servlet
1.1</TITLE></HEAD> <BODY>\n");
    html.append("<b>Java for Bioinformatics: </b>");
    html.append("<font color=red><b>PubMed Servlet version
1.1</b></font>\n");
    html.append("<BR><BR><B>Please enter a term to search
on NCBI PubMed:</B> <BR><BR>\n");
    html.append("<FORM METHOD=GET>\n");
    html.append("<INPUT TYPE=TEXT
NAME=searchTerm><BR><BR>\n");
    html.append("<INPUT TYPE=SUBMIT VALUE=\"Search
PubMed\"><BR>\n");
    html.append("</FORM>\n");

    String term = req.getParameter("searchTerm");
    if (term != null) {
      html.append("<BR><HR><BR>");
      html.append("You have searched NCBI for the term
<font color=red>'" + term + "'</font>.");
      URL url = new
URL("http://www.ncbi.nlm.nih.gov/entrez/query.fcgi?dispmax=10
&db=pubmed&cmd=search&term=" + URLEncoder.encode(term, "UTF-
8"));
      URLConnection urlConnection = url.openConnection();
      BufferedReader reader = new BufferedReader(new
InputStreamReader(urlConnection.getInputStream()));

      String s = null;
```

```
        RE pmidRE = new RE("PMID: ([0-9]+) \\[PubMed");
        Collection pmids = new ArrayList();

        while ((s = reader.readLine()) != null) {
          if (pmidRE.match(s)) {
            pmids.add(pmidRE.getParen(1));
          }
        }
        reader.close();

        html.append("<BR><br>PMIDs found:<br>\n");
        int i = 1;

        for (Iterator iterator = pmids.iterator();
iterator.hasNext();) {
          String s1 = (String) iterator.next();
          html.append("<a href=\"")

.append("http://www.ncbi.nlm.nih.gov/entrez/query.fcgi?cmd=Re
trieve&db=pubmed&dopt=Abstract&list_uids=")
            .append(s1)
            .append("\">")
            .append(s1)
            .append("</a>\n");
          if (iterator.hasNext() && i++ != 5) {
            html.append(" - ");
          } else {
            html.append("<BR>");
          }
        }
      }
      html.append("</BODY></HTML>\n");
      out.print(html.toString());
    }
  }
```

The next few lines of code iterate over the array for each of the PMIDs of abstracts matching the search term and print them out along with a hyperlink to the original abstract on PubMed. The structure of the servlet and its component files is shown below (**Fig. 4.9**).

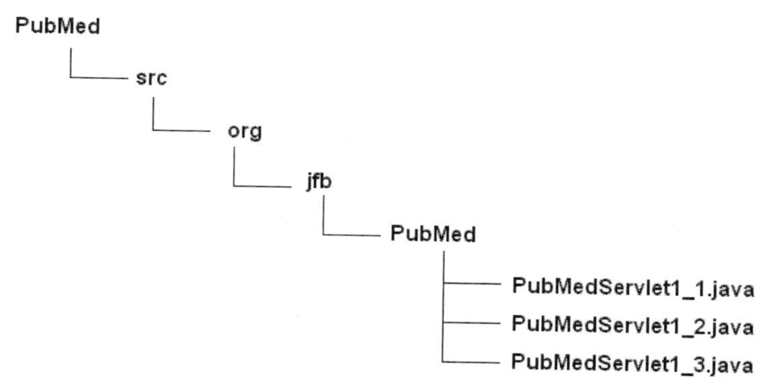

Fig. 4.9. The PubMed servlet structure

To see the *servlet* in action, start the Apache Tomcat Server, compile the code and run it with the command:

```
ant install
```

Apache Ant is a Java-based build tool used to manage the different steps in the development cycle of an application, which include compilation of the code libraries needed for the application, creating the necessary JARs for deploying an application, etc. It is available from The Apache Software Foundation website. For further information on installation and use, please refer to the Appendix.

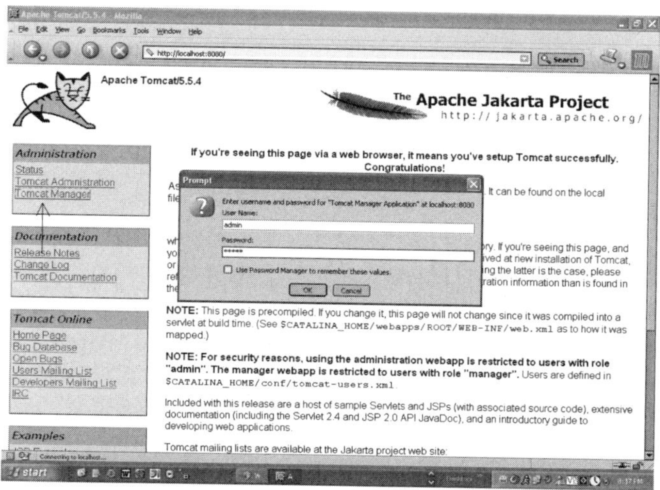

Fig. 4.10. Logging into the Tomcat Manager

Open the following URL:

```
http://localhost:8080
```

When the Apache Tomcat welcome page loads, click on the Tomcat Manager visible on the left panel and login into the server using the credentials you specified during installation (**Fig. 4.10**). Access the latest build of the application to view the servlet. The output of the search with the keyword HIV using the first version of our program, which we will call PubMed Servlet version 1.1, is shown in **Fig. 4.11**.

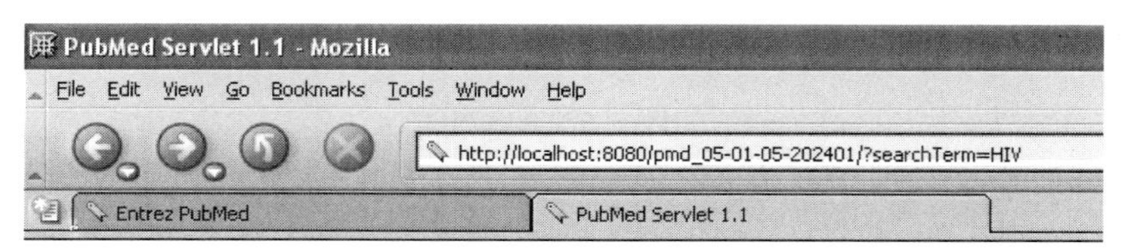

Java for Bioinformatics: PubMed Servlet version 1.1

Please enter a term to search on NCBI PubMed:

Search PubMed

You have searched NCBI for the term 'HIV'.

PMIDs found:
15630704 - 15630678 - 15630469 - 15630452 - 15630446
15630430 - 15630360 - 15629958 - 15629857 - 15629784

Fig. 4.11. Output from the PubMed servlet using search term "HIV"

In the first version of the application, we are simply validating our approach and displaying just the PMIDs for the abstracts that match the entered keyword. To check that the code works and retrieves the correct data, we hyperlink the PMIDs to the original abstracts on PubMed. Clicking on 15630704, for example, opens up the abstract corresponding to the PMID for the abstract that shows up in the search performed directly on the NCBI PubMed webpage (**Fig. 4.12**).

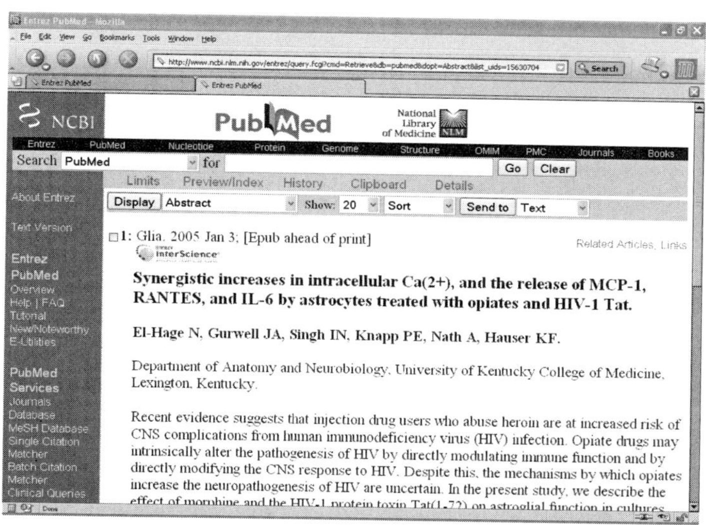

Fig. 4.12. PubMed article corresponding to PMID 15630704

The results in **Fig. 4.11** and **Fig. 4.12** are identical to the search output obtained from a search with the keyword 'HIV' at NCBI PubMed at the time of this writing (**Fig. 4.13**).

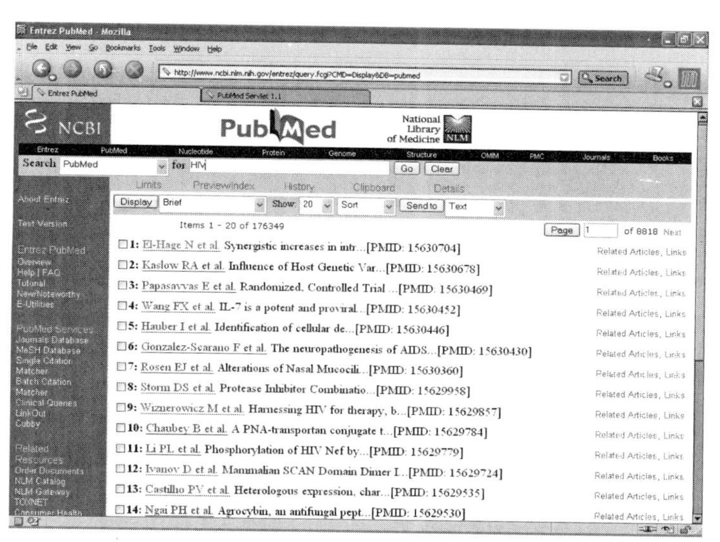

Fig. 4.13. Results of NCBI PubMed search with keyword "HIV"

Displaying PubMed Abstracts

In order to make the search output more useful for researchers, we would like to parse the abstract from each citation and make it available for viewing right up front as part of the search results. We will now create the code to parse out the abstract from each of the articles that are returned by a search.

The general framework of the program is as follows:

1. Create the search form
2. Retrieve the keyword(s) provided by the user
3. Retrieve PMIDs from PubMed corresponding to the search term
4. Retrieve abstracts based on each of the PMIDs obtained in step 1
5. Iterate #4 until all abstracts have been retrieved

To create the search form, we create a method called `createSearchForm()` which creates a variable of type *StringBuffer* called html and append the various html tags to it in succession:

```
    private StringBuffer createSearchForm() {
      StringBuffer html=new StringBuffer();
      html.append("<HTML>");
      html.append("<HEAD><TITLE>" + TITLE +
"</TITLE></HEAD><BODY>");
      html.append("<b>Java for Bioinformatics: </b>");
      //html.append("<font color=red><b>PubMed Servlet
version 1.1</b></font>\n");

      html.append("<font color=red><h1>" + TITLE +
"</h1></font>");
      //html.append("<B>Please enter a userKeywords to search
on NCBI:</B><BR><BR>\n");
      html.append("<BR><B>Please enter a term to search on
NCBI PubMed: </B><BR><BR>\n");

      html.append("<FORM METHOD=GET>\n");
      html.append("<INPUT TYPE=TEXT NAME=" + KEYWORDS +
"><BR><BR>\n");
      html.append("<INPUT TYPE=SUBMIT VALUE=\"Search
PubMed\"><BR>\n");
      html.append("</FORM>\n");
      return html;
    }
```

Note that the text box for entering keywords is called KEYWORDS. We will use this name to retrieve the user-supplied keywords. The resulting search form for the next iteration of the application, which we will call PubMed Servlet version 1.2, is shown in **Fig. 4.13**.

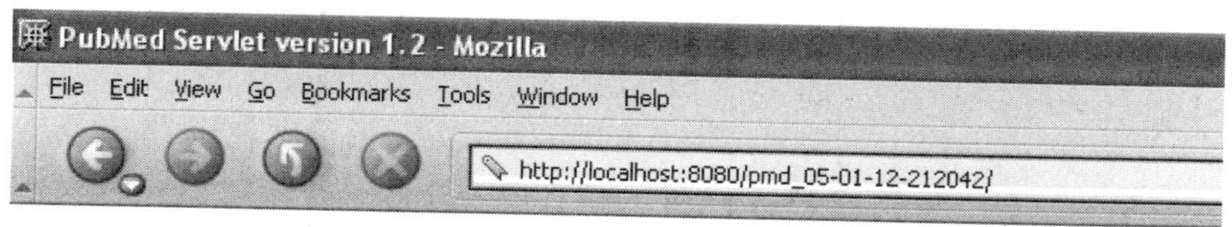

Java for Bioinformatics:

PubMed Servlet version 1.2

Please enter a term to search on NCBI PubMed:

Search PubMed

Build #05-01-12-212042

Fig. 4.13. PubMed servlet search form version 1.2

We then retrieve the keyword(s) from the search box using a method called `getUserKeywords()`:

```
String userKeywords = getUserKeywords(req);
```

This method takes in the *HttpServletRequest* req object as a parameter to return the keywords:

```
private String getUserKeywords(HttpServletRequest req) {
    return req.getParameter(KEYWORDS);
}
```

The next few lines perform some basic user input validation. If you press the search button without supplying any keywords, for example, the program will return an error message: "Please enter keywords to search." (**Fig. 4.14**).

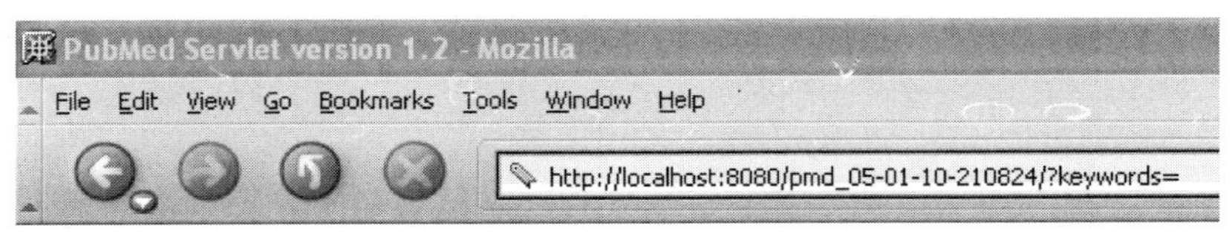

Fig. 4.14. User-input validation

We then create a variable of type *StringBuffer* called `sbPmids` to store PMIDs corresponding to the search terms and a *String* variable called `searchURL` to specify the search URL:

```
  StringBuffer sbPmids = null;
       final String searchURL =
"http://www.ncbi.nlm.nih.gov/entrez/query.fcgi?dispmax=10&db=
pubmed&cmd=search&term=" + URLEncoder.encode(userKeywords,
"UTF-8");
```

Next we write a method called `getPmids()` to retrieve PMIDs from the keywords. The method takes one parameter, the `searchURL`, which in turns contains the keyword(s) embedded in it. The result of the operation is stored in an object called `sbPmids`:

```
sbPmids = getPmids(searchURL);
```

We place the method within a *try-catch block* we had briefly mentioned earlier to catch any *exceptions* that may arise while the request is sent to PubMed. If we do indeed encounter an exception, the method will trap the error, print out the offending error message and exit.

```
try {
  sbPmids = getPmids(searchURL);
}
catch (IOException ioe) {
  ioe.printStackTrace();
  errorMes = "<BR><BR><font color=red>We are sorry, the
system could not establish connection to the NCBI PubMed
server " + "with the URL &quot;" + searchURL + "&quot;.
Please try again later.</font><BR><BR>";
  }
```

The method `getPmids()` itself looks like this:

```
private StringBuffer getPmids(String searchURL) throws
IOException {
    BufferedReader reader = new BufferedReader(new
InputStreamReader(new
URL(searchURL).openConnection().getInputStream()));
    StringBuffer sbPmids = new StringBuffer();
    String pmid;
    String s = null;

    while ((s = reader.readLine()) != null) {
      if (pmidRE.match(s)) {
        pmid = pmidRE.getParen(1);
        sbPmids.append(pmid + ",");
      }
    }
    reader.close();
    final int length = sbPmids.length();

    if (length > 0) {
      sbPmids.delete(length - 1, length);
      return sbPmids;
    } else {
      return null;
    }
  }
```

The method:

```
  BufferedReader reader = new BufferedReader(new
InputStreamReader(new
URL(searchURL).openConnection().getInputStream()));
```

can be broken down into more readable chunks of code as follows:

```
  URLConnection urlConnection = new
URL(searchURL).openConnection();
  InputStream inputStream = urlConnection.getInputStream();
  BufferedReader reader = new BufferedReader(new
InputStreamReader(inputStream));
```

If no exceptions have been raised and if PMIDs have been obtained as a result of the search, we proceed to get the abstracts from the PMIDs. The method we use here is called `getAbstracts()` and returns an object of type *StringBuffer* called abstracts. The method takes a parameter called `urlAddress`, which specifies the location of the abstract based on the corresponding PMID:

```
        if (errorMes == null) {
            if (sbPmids != null) {
              String urlAddress = citationString +
URLEncoder.encode(sbPmids.toString(),
                  "UTF-8");
              StringBuffer abstracts = null;

             // 3. Retrieve the abstracts from the PubMed IDs
              try {
                abstracts = getAbstracts(urlAddress);
              } catch (IOException ioe) {
                ioe.printStackTrace();
                errorMes = "<BR><BR><font color=red>We are
sorry, the system could not retrieve the abstracts using
keyword(s) &quot;"
                    + userKeywords + "&quot; with the URL
<PRE>&quot;" + urlAddress + "&quot;<PRE></font><BR><BR>";
              }
```

The code for the method `getAbstracts()` is as follows:

```
  private StringBuffer getAbstracts(String urlAddress) throws
IOException {
      BufferedReader citationReader =
        new BufferedReader(new InputStreamReader(new
URL(urlAddress).openConnection().getInputStream()));
      StringBuffer abstracts = new StringBuffer();
```

```
    String s;
    while ((s = citationReader.readLine()) != null) {
      abstracts.append(s);
    }
    return abstracts;
  }
```

Next we get information from the matching articles corresponding to each abstract. This includes information such as the title, authors, source journal in which the article was published, and the like. An example of the MEDLINE format, which is parsed to extract this information, is shown in **Fig. 4.15**. Note the tags on the left – PMID, OWN, DP, TI, AB, AU, AD, SO. These represent respectively the PubMed ID, the owner (the organization that supplied the citation data for MEDLINE), date of publication, title, abstract, authors, address and source journal.

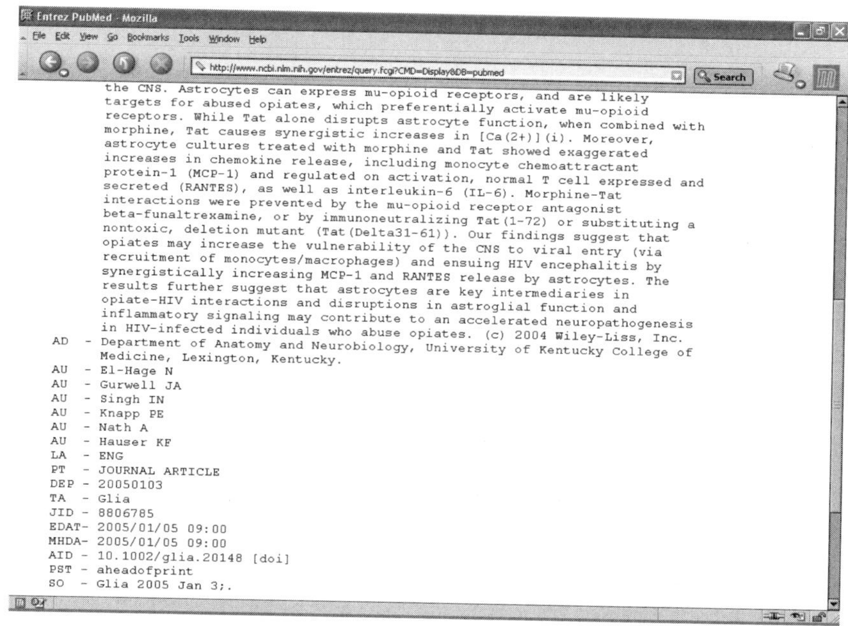

Fig. 4.15. The MEDLINE format

Parsing of these elements is done using the jakarta regular expression library. Let's see how we can parse the PMID from the MEDLINE record displayed above. Note that the PMID is bounded by the tags PMID and OWN as shown in the enlarged **Fig. 4.16** below.

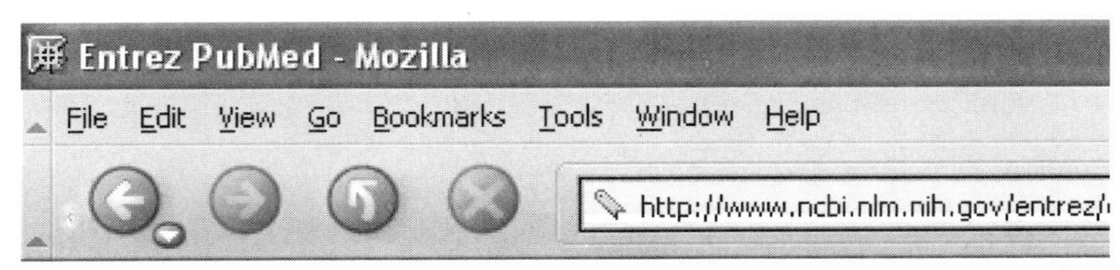

☐ **1:** El-Hage N et al. Synergistic increases in

```
PMID- 15630704
OWN - NLM
STAT- Publisher
DA  - 20050104
PUBM- Print-Electronic
IS  - 0894-1491
DP  - 2005 Jan 3
```

Fig. 4.16. Parsing the PMID

We could use regular expressions to capture the PMID and other information if all the MEDLINE records had the same standard format. A few of these tags are not present wherever information is not available. For example, sometimes the abstract is not available. In such cases the AB tag is not present in the MEDLINE record which makes it a little more difficult to construct a regular expression that is generic enough for all cases. We demonstrate an alternate method that locates the position of each start and end tag and captures everything in between. We will declare the tags we will use to construct regular expressions at the beginning of the program:

```
private static final String pmidTag = "PMID- ";
private static final String pmidEndTag = "OWN - ";
private static final String titleStartTag = "TI  - ";
private static final String titleEndTag = "PG  - ";
private static final String abstractTag = "AB  - ";
private static final String abstractEndTag = "AD  - ";
private static final String fauthorStartTag = "FAU - ";
private static final String authorStartTag = "AU  - ";
private static final String authorEndTag = "LA  - ";
private static final String srcTag = "SO  -";
private static final String medlineEndTag = "</pre>";
```

We will next create code for the method that we will call `getArticleInfo()` for retrieving the information:

```
   private StringBuffer getArticleInfo(StringBuffer tmp, int
pmidStart, int endMedline) {
        StringBuffer articleTmp = new StringBuffer();
        String pmid1 = tmp.substring(pmidStart +
pmidTag.length(), tmp.indexOf(pmidEndTag));

        int titleStart = tmp.indexOf(titleStartTag);
        int titleEnd = tmp.indexOf(titleEndTag);

        if (titleEnd < 0 || titleEnd > endMedline)
          titleEnd = tmp.indexOf(abstractTag);

        int abstractStart = tmp.indexOf(abstractTag);

        if (titleEnd < 0 || titleEnd > endMedline) {
          titleEnd = tmp.indexOf(fauthorStartTag);
        }

        if (titleEnd < 0 || titleEnd > endMedline) {
          titleEnd = tmp.indexOf(fauthorStartTag);
        }
        String title = null;

        if (0 <= titleStart && titleStart < endMedline) {
          titleStart += titleStartTag.length();
          title = tmp.substring(titleStart,
titleEnd).replaceAll("(\\s+)", " ");
        }

        int end = tmp.indexOf(abstractEndTag);
        String tmpAbstractTag = abstractEndTag;

        if (end < 0 || end > endMedline) {
          end = tmp.indexOf(fauthorStartTag);
          tmpAbstractTag = fauthorStartTag;

          if (end < 0 || end > endMedline) {
            end = tmp.indexOf(authorEndTag);
            tmpAbstractTag = authorStartTag;
          }
        }

        String article = null;
        if (abstractStart + tmpAbstractTag.length() <= end) {
          article = tmp.substring(abstractStart +
tmpAbstractTag.length(), end).replaceAll("(\\s+)", " ");
        }

        int authorStart = tmp.indexOf(authorStartTag);
        String authors = null;

        if (0 <= authorStart && authorStart < endMedline) {
          authorStart += authorStartTag.length();
```

```
          int authorEnd = tmp.indexOf(authorEndTag);
          authors = tmp.substring(authorStart,
authorEnd).replaceAll(authorStartTag, ",
").replaceAll(fauthorStartTag, ", ");
        }

        int srcStart = tmp.indexOf(srcTag);
        String journal = null;
        if (0 <= srcStart && srcStart < endMedline) {
          journal = tmp.substring(srcStart + srcTag.length(),
endMedline);
        }

        // Let's create the document
        articleTmp.append("<a href=\"" + PUBMED_ARTICLE_LK +
pmid1 + "\">" + pmid1 + "</a>").append("<BR>");
        articleTmp.append("<U>Journal</u>: ");
        articleTmp.append(journal != null ? journal : "No
journal listed").append("<BR>");
        articleTmp.append("<u>Authors</u>: ");
        articleTmp.append(authors != null ? authors : "No
authors listed").append("<BR>");
        articleTmp.append("<u>Title</u>: ");
        articleTmp.append(title != null ? title : "No
title").append("<BR>");
        articleTmp.append("<u>Abstract</u>: ");
        articleTmp.append(article != null ? article : "No
article").append("<BR>");
        return articleTmp;
      }
```

The output of the second version of the PubMed servlet program that automatically parses the abstracts for each of the returned citations is shown in **Fig. 4.17**. Each of the abstracts is marked at the beginning with the PubMed ID which in turn is hyperlinked to the citation on PubMed if the user wishes to see the original record at NCBI.

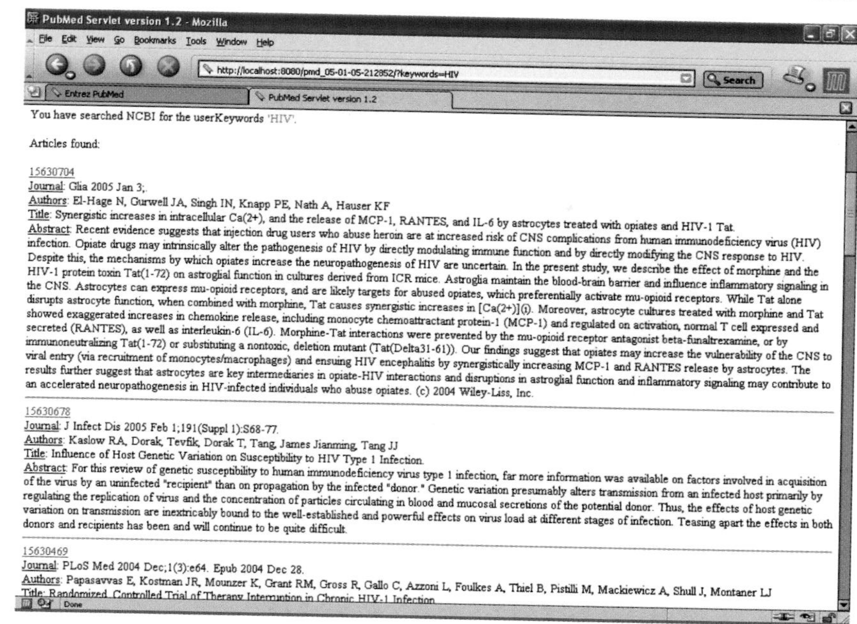

Fig. 4.17. Displaying abstracts for matching PubMed articles

The complete code for the second version of PubMed servlet (version 1.2) is shown in in **Listing 4.2**.

Listing 4.2. PubMed Servlet version 1.2

```
package org.jfb.PubMed;

import org.apache.regexp.RE;

import javax.servlet.ServletException;
import javax.servlet.http.HttpServlet;
import javax.servlet.http.HttpServletRequest;
import javax.servlet.http.HttpServletResponse;
import java.io.*;
import java.net.URL;
import java.net.URLEncoder;
import java.net.URLConnection;
import java.util.Properties;

public class PubMedServlet1_2 extends HttpServlet {
   private static final String TITLE = "PubMed Servlet
version 1.2";
   private static final String KEYWORDS = "keywords";
   private static final String PUBMED_ARTICLE_LK =
```

```
"http://www.ncbi.nlm.nih.gov/entrez/query.fcgi?cmd=Retrieve&d
b=pubmed&dopt=Abstract&list_uids=";
    private static final String citationString =

"http://www.ncbi.nlm.nih.gov/entrez/query.fcgi?cmd=Retrieve&d
b=PubMed&dopt=medline&list_uids=";
    private static final RE pmidRE = new RE("PMID: ([0-9]+)
\\[PubMed");

    private static final String pmidTag = "PMID- ";
    private static final String pmidEndTag = "OWN - ";
    private static final String titleStartTag = "TI  - ";
    private static final String titleEndTag = "PG  - ";
    private static final String abstractTag = "AB  - ";
    private static final String abstractEndTag = "AD - ";
    private static final String firstAuthorStartTag = "FAU -
";
    private static final String authorStartTag = "AU  - ";
    private static final String authorEndTag = "LA  - ";
    private static final String srcTag = "SO  -";
    private static final String medlineEndTag = "</pre>";

    protected void doGet(HttpServletRequest req,
HttpServletResponse res)
      throws ServletException, IOException {
      StringBuffer html = new
StringBuffer(createSearchForm());

      // 1. Extract the user-supplied keywords
      String userKeywords = getUserKeywords(req);
      if (userKeywords != null) {
        if (userKeywords.equals("")) {
          String errorMes;
          errorMes = "<BR><BR><font
color=red><b>ERROR</b><BR>Please enter keywords to
search!</font><Br><BR>";
          html.append(errorMes);
        } else {
          html.append("<BR><HR><BR>");
          html.append("You have searched NCBI PubMed with the
keywords <font color=red>'" + userKeywords + "'</font>.");

          // 2. Retrieve the PubMed IDs from the user
  // keywords
          StringBuffer sbPmids = null;   //sbpmids
          final String searchURL =
"http://www.ncbi.nlm.nih.gov/entrez/query.fcgi?dispmax=10&db=
pubmed&cmd=search&term="
            + URLEncoder.encode(userKeywords, "UTF-8");

          String errorMes = null;
```

```
            try {
  //           if (true) throw new IOException("Testing the
connection failure here!");
              sbPmids = getPmids(searchURL);
            } catch (IOException ioe) {
              ioe.printStackTrace();
              errorMes = "<BR><BR><font color=red>We are sorry,
the system could not establish connection to the NCBI PubMed
server "
                + "with the URL &quot;" + searchURL + "&quot;.
Please try again later.</font><BR><BR>";
            }

            if (errorMes == null) {
              if (sbPmids != null) {
                String urlAddress = citationString +
URLEncoder.encode(sbPmids.toString(), "UTF-8");
                StringBuffer abstracts = null;

                // 3. Retrieve the abstracts from the PubMed
  // IDs
                try {
                  abstracts = getAbstracts(urlAddress);
                } catch (IOException ioe) {
                  ioe.printStackTrace();
                  errorMes = "<BR><BR><font color=red>We are
sorry, the system could not retrieve the abstracts using
keyword(s) &quot;"
                    + userKeywords + "&quot; with the URL
<PRE>&quot;" + urlAddress + "&quot;<PRE></font><BR><BR>";
                }

                if (errorMes == null) {
                  int pmidStart = abstracts.indexOf(pmidTag);
                  StringBuffer tmp = abstracts;
                  html.append("<BR><br>Articles
found:<br><BR>\n");
                  StringBuffer article;

                  // 4. Extract information from the articles
                  try {
                    while (pmidStart != -1) {
                      int endMedline =
tmp.indexOf(medlineEndTag);
                      article = getArticleInfo(tmp, pmidStart,
endMedline);

                      html.append(article);

                      tmp.delete(0, endMedline +
medlineEndTag.length());
                      pmidStart = tmp.indexOf(pmidTag);

                      if (pmidStart != -1) {
```

```
                        html.append("<HR>");
                      }
                    }
                  } catch (Exception e) {
                    e.printStackTrace();
                    errorMes = "<BR><BR><font
color=red><h1>ERROR</h1><BR>We are sorry, the system could
not retrieve the articles for PMIDs <PRE>&quot;"
                      + sbPmids + "&quot;<PRE></font><BR><BR>";
                    html.append(errorMes);
                  }
                } else {
                  html.append(errorMes);
                }
              } else {
                html.append("<BR>No abstracts found!");
              }
            } else {
              html.append(errorMes);
            }
          }
        }

        appendBuildProperty(html);
        html.append("</BODY></HTML>\n");

        // 5. Print the results
        res.setContentType("text/html");
        PrintWriter out = res.getWriter();
        out.print(html);
      }

      private String getUserKeywords(HttpServletRequest req) {
        return req.getParameter(KEYWORDS);
      }

      private StringBuffer createSearchForm() {
        StringBuffer html=new StringBuffer();
        html.append("<HTML>");
        html.append("<HEAD><TITLE>" + TITLE +
"</TITLE></HEAD><BODY>");
        html.append("<b>Java for Bioinformatics: </b>");

        html.append("<font color=red><h1>" + TITLE +
"</h1></font>");
            html.append("<BR><B>Please enter a term to search
on NCBI PubMed: </B><BR><BR>\n");

        html.append("<FORM METHOD=GET>\n");
        html.append("<INPUT TYPE=TEXT NAME=" + KEYWORDS +
"><BR><BR>\n");
        html.append("<INPUT TYPE=SUBMIT VALUE=\"Search
PubMed\"><BR>\n");
```

```
        html.append("</FORM>\n");
        return html;
    }

    private StringBuffer getPmids(String searchURL) throws
IOException {
        URLConnection urlConnection = new
URL(searchURL).openConnection();
        InputStream inputStream =
urlConnection.getInputStream();
        BufferedReader reader = new BufferedReader(new
InputStreamReader(inputStream));

        StringBuffer sbPmids = new StringBuffer();
        String pmid;
        String s = null;

        while ((s = reader.readLine()) != null) {
          if (pmidRE.match(s)) {
            pmid = pmidRE.getParen(1);
            sbPmids.append(pmid + ",");
          }
        }
        reader.close();
        final int length = sbPmids.length();

        if (length > 0) {
          sbPmids.delete(length - 1, length);
          return sbPmids;
        } else {
          return null;
        }
    }

    private StringBuffer getAbstracts(String urlAddress)
throws IOException {
        BufferedReader citationReader =
          new BufferedReader(new InputStreamReader(new
URL(urlAddress).openConnection().getInputStream()));
        StringBuffer abstracts = new StringBuffer();

        String s;
        while ((s = citationReader.readLine()) != null) {
          abstracts.append(s);
        }
        return abstracts;
    }

    private StringBuffer getArticleInfo(StringBuffer tmp, int
pmidStart, int endMedline) {
        StringBuffer articleTmp = new StringBuffer();
        String pmid1 = tmp.substring(pmidStart +
pmidTag.length(), tmp.indexOf(pmidEndTag));
```

```
      int titleStart = tmp.indexOf(titleStartTag);
      int titleEnd = tmp.indexOf(titleEndTag);

      if (titleEnd < 0 || titleEnd > endMedline)
        titleEnd = tmp.indexOf(abstractTag);

      int abstractStart = tmp.indexOf(abstractTag);

      if (titleEnd < 0 || titleEnd > endMedline) {
        titleEnd = tmp.indexOf(firstAuthorStartTag);
      }
      String title = null;

      if (0 <= titleStart && titleStart < endMedline) {
        titleStart += titleStartTag.length();
        title = tmp.substring(titleStart,
titleEnd).replaceAll("(\\s+)", " ");
      }

      int end = tmp.indexOf(abstractEndTag);
      String tmpAbstractTag = abstractEndTag;

      if (end < 0 || end > endMedline) {
        end = tmp.indexOf(firstAuthorStartTag);
        tmpAbstractTag = firstAuthorStartTag;

        if (end < 0 || end > endMedline) {
          end = tmp.indexOf(authorEndTag);
          tmpAbstractTag = authorStartTag;
        }
      }

      String article = null;
      if (abstractStart + tmpAbstractTag.length() <= end) {
        article = tmp.substring(abstractStart +
tmpAbstractTag.length(), end).replaceAll("(\\s+)", " ");
      }

      int authorStart = tmp.indexOf(authorStartTag);
      String authors = null;

      if (0 <= authorStart && authorStart < endMedline) {
        authorStart += authorStartTag.length();
        int authorEnd = tmp.indexOf(authorEndTag);
        authors = tmp.substring(authorStart,
authorEnd).replaceAll(authorStartTag, ",
").replaceAll(firstAuthorStartTag, ", ");
      }

      int srcStart = tmp.indexOf(srcTag);
      String journal = null;
      if (0 <= srcStart && srcStart < endMedline) {
```

```
        journal = tmp.substring(srcStart + srcTag.length(),
endMedline);
      }

      // Let's create the document
      articleTmp.append("<a href=\"" + PUBMED_ARTICLE_LK +
pmid1 + "\">" + pmid1 + "</a>").append("<BR>");
      articleTmp.append("<U>Journal</u>: ");
      articleTmp.append(journal != null ? journal : "No
journal listed").append("<BR>");
      articleTmp.append("<u>Authors</u>: ");
      articleTmp.append(authors != null ? authors : "No
authors listed").append("<BR>");
      articleTmp.append("<u>Title</u>: ");
      articleTmp.append(title != null ? title : "No
title").append("<BR>");
      articleTmp.append("<u>Abstract</u>: ");
      articleTmp.append(article != null ? article : "No
article").append("<BR>");
      return articleTmp;
    }

    private void appendBuildProperty(StringBuffer html) {
      Properties buildInfo = null;

      try {
        buildInfo = new Properties();
        InputStream buildStream =
getClass().getClassLoader().getResourceAsStream("/build-
info.txt");
        buildInfo.load(buildStream);
      } catch (Throwable e) {
        e.printStackTrace();
      }

      if (buildInfo != null) {
        html.append("<BR><HR><font color=grey>Build #");
        html.append(buildInfo.getProperty("buildNumber"));
        html.append("</font>\n");
      }
    }

    public static void main(String[] args) throws Exception {
      new PubMedServlet1_2();
    }
  }
```

Highlighting Search Terms in Retrieved Abstracts

In version 1.3 of the PubMed servlet, we will enhance the usefulness of the search results by highlighting the search terms in the retrieved abstracts. One way to do this is to convert the search terms and the abstract into lower case, locate the matches and then highlight the terms in the abstract. In this method, we lose the case of the words in the original abstract (because we converted that into lower case). To fix this, we could find the exact location of the match and the length of the match and use the original abstract to highlight the matching term(s).

Another way is to use the `equalsIgnoreCase()` method which compares strings irrespective of case. For example, the following code will find a match to the term "HIV" in text even if it contains HIV in different forms such as hiv, Hiv, HIv, hIV, etc.

```
if (word.equalsIgnoreCase( "HIV" ) ) {
  //code for highlighting matching terms;
}
```

To use this method, we have to first create an array of words in the abstract and test if any of the individual words match the search term. However, there are limitations to this method also. The `equalsIgnoreCase()` method searches for exact matches and will not find words containing punctuation marks and other characters. If, for example, HIV-1 is found at the end of a sentence, the array element will be "HIV." (with a period) and "HIV" is not equal to "HIV.". To fix this we need to get rid of all such punctuation marks and other special characters.

An easier method to circumvent these issues is described below. In this method, we iterate over the text in the abstract highlighting each term as it is found. The regular expression itself is of the type:

```
(a|A) (b|B) (c|C)…
```

which will match any word irrespective of case. Surrounding such expression in parentheses allows us to extract specific sub-strings from a string based on a specified pattern. This is implemented in code as follows:

```
  StringBuffer sb = new StringBuffer("(");
  for (char c : chars) {
      char charUp = Character.toUpperCase(c);
      char charLo = Character.toLowerCase(c);

sb.append("(").append(charLo).append("|").append(charUp).appe
nd(")");
  }
  sb.append(")");
```

```
  final String regex = "^" + sb.toString() + "|[^a-zA-Z]" +
sb.toString();
```

We will not only highlight the search term in the abstracts, we will also color them differently for better visibility and readability. To do this, we need to declare an array called COLOR of color elements to store the selection of colors we wish to use:

```
  private static final String[] COLOR = new String[]{"blue",
"#98cc02", "purple", "red", "#f7dc88"};
```

For each of the characters in the search term, a regular expression of the type indicated above (with both lower and upper case forms) is created. Next when the term is found in the article text, it is highlighted using a different color for each matching term.

```
  highlightedText = re.subst(highlightedText, "\\\\<b><font
style=\"\\\\+2\" color=\"" + COLOR[i] + "\">$0</font></b>",
                        RE.REPLACE_BACKREFERENCES);
          }
```

The complete code for the method which we will call `highlight()` is as follows:

```
  private String highlight(String articleText, String[]
terms) {
          String highlightedText = new String(articleText);
          for (int i = 0; i < terms.length; i++) {
              final String term = terms[i];
              final char[] chars = term.toCharArray();

  // Here we are creating the regular expression to find any
// word irrespective of case.
              StringBuffer sb = new StringBuffer("(");
              for (char c : chars) {
                char charUp = Character.toUpperCase(c);
                char charLo = Character.toLowerCase(c);

sb.append("(").append(charLo).append("|").append(charUp).appe
nd(")");
              }
              sb.append(")");
              final String regex = "^" + sb.toString() +
"|[^a-zA-Z]" + sb.toString();

              // Replace the text by a HTML FONT tag that
  // wraps the term found
              RE re = new RE(regex);
```

```
                highlightedText = re.subst(highlightedText,
"\\\\<b><font style=\"\\\\+2\" color=\"" + COLOR[i] +
"\">$0</font></b>", RE.REPLACE_BACKREFERENCES);
            }
            return highlightedText;
        }
```

The regular expression for highlighting matched text with colored text is constructed using the Jakarta regular expression library. In particular we are using the *subst* method (short for substring), which is defined as follows:

```
re.subst(string1, string2, rules)
```

where,

`string1`: the *String* to make the substitution in
`string2`: *String* to substitute into `string1`
`rules`: rules that define how substitutions are to be done in `string1`

To refer to the contents of a parenthesized expression within a regular expression, we use what are known as '*backreferences*'. The first backreference in a regular expression is denoted by \1, the second by \2 and so on.

The rules are set as follows:

REPLACE_FIRSTONLY: replace only the first occurrence of the regular expression in string1

REPLACE_ALL: replace all occurrences of the regular expression in string1

REPLACE_BACKREFERENCES: all backreferences will be processed, which in this case means that all matched patterns within the article text will be replaced with string2

In our case,

`string1` = highlightedText
`string2` = "\\\\<b><font style=\"\\\\+2\" color=\"" + COLOR[i] + "\">$0</font></b>"
and

```
rules = RE.REPLACE_BACKREFERENCES
```

The extra backslashes in `string2` are *escape characters*. Note that the expression "`$0`" represents the whole match, which in this case, represent the search term(s). The output of PubMed servlet version 1.3 obtained from an ANDed search of the terms HIV AND AIDS is shown in **Fig. 4.18**.

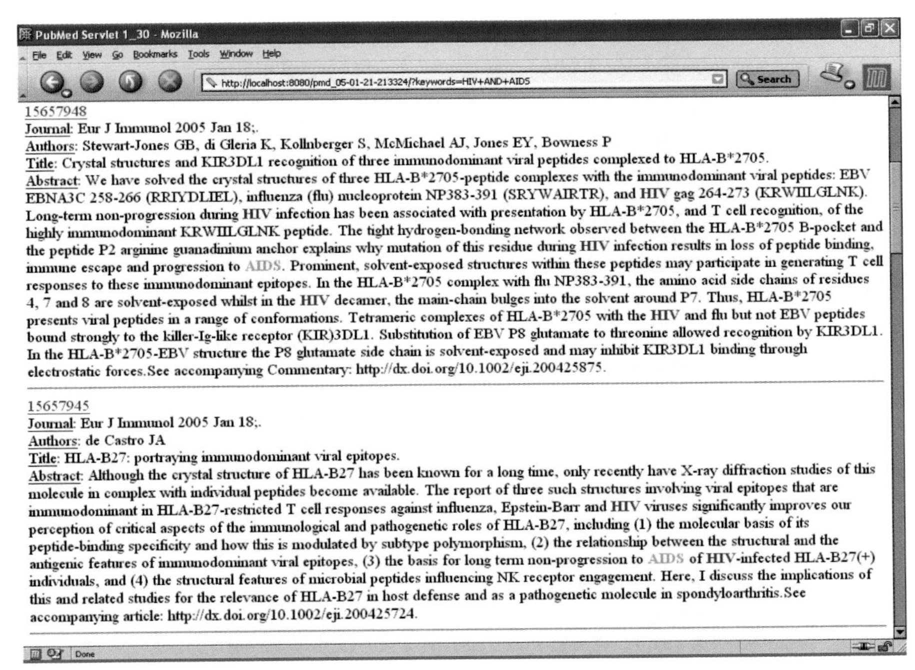

Fig. 4.18. Highlighting search terms in PubMed abstracts

As the output shows, both keywords have been highlighted (blue and green respectively, as specified in the array of HTML colors).

The complete code for PubMed servlet version 1.3 is shown in **Listing 4.3**.

Listing 4.3. PubMed Servlet version 1.3

```
package org.jfb.PubMed;

import org.apache.regexp.RE;
import javax.servlet.ServletException;
import javax.servlet.http.HttpServlet;
import javax.servlet.http.HttpServletRequest;
```

```
import javax.servlet.http.HttpServletResponse;
import java.io.*;
import java.net.URL;
import java.net.URLEncoder;
import java.util.Properties;

public class PubMedServlet1_3 extends HttpServlet {
    private static final String TITLE = "PubMed Servlet
1_30";
    private static final String KEYWORDS = "keywords";
    private static final String PUBMED_ARTICLE_LK =

"http://www.ncbi.nlm.nih.gov/entrez/query.fcgi?cmd=Retrieve&d
b=pubmed&dopt=Abstract&list_uids=";
    private static final String citString =

"http://www.ncbi.nlm.nih.gov/entrez/query.fcgi?cmd=Retrieve&d
b=PubMed&dopt=medline&list_uids=";
    private static final RE pmidRE = new RE("PMID: ([0-9]+)
\\[PubMed");

    private static final String pmidTag = "PMID- ";
    private static final String pmidEndTag = "OWN - ";
    private static final String titleStartTag = "TI  - ";
    private static final String titleEndTag = "PG  - ";
    private static final String abstractTag = "AB  - ";
    private static final String abstractEndTag = "AD  - ";
    private static final String firstAuthorStartTag = "FAU
- ";
    private static final String authorStartTag = "AU  - ";
    private static final String authorEndTag = "LA  - ";
    private static final String srcTag = "SO  -";
    private static final String medlineEndTag = "</pre>";
    private static final String[] COLOR = new
String[]{"blue", "#98cc02", "purple", "red", "#f7dc88"};
    private String[] params;

    protected void doGet(HttpServletRequest req,
HttpServletResponse res) throws ServletException, IOException
{
        StringBuffer html = new StringBuffer();

        // 1. Retrieve the user supplied keywords
        printHeader(html);
        String userKeywords = req.getParameter(KEYWORDS);

        if (userKeywords != null) {
            params =
userKeywords.replaceAll("\\s*(\\+|((a|A)(N|n)(D|d))|((o|O)(r|
R)))\\s*", " ").split(" ");
            html.append("<BR><HR><BR>");
            html.append("You have searched NCBI for the
userKeywords '"
```

```
                              + highlight(userKeywords, this.params)
+ "'.");

                // 2. Retrieve the PubMed IDs from abstracts
   // matching user supplied keywords.
                StringBuffer sbPmids = null;
                final String spec =
"http://www.ncbi.nlm.nih.gov/entrez/query.fcgi?dispmax=10&db=
pubmed&cmd=search&term="
                              + URLEncoder.encode(userKeywords, "UTF-
8");

                String errorMes = null;
                System.out.println("spec = " + spec);
                try {
                    sbPmids = getPmids(spec);
                } catch (IOException ioe) {
                    ioe.printStackTrace();
                    errorMes = "<BR><BR><font color=red>We are
sorry, the system could not retrieve the PubMed IDs using
keyword(s) &quot;"
                                  + userKeywords + "&quot; with the
URL <PRE>&quot;" + spec + "&quot;<PRE></font><BR><BR>";
                }

                if (errorMes == null) {
                    if (sbPmids != null) {
                        String urlAddress = citString +
URLEncoder.encode(sbPmids.toString(), "UTF-8");
                        StringBuffer abstracts = null;

                        // 3. Retrieve abstracts corresponding
// to the PubMed IDs
                        try {
                            abstracts =
getAbstracts(urlAddress);
                        } catch (IOException ioe) {
                            ioe.printStackTrace();
                            errorMes = "<BR><BR><font
color=red>We are sorry, the system could not retrieve the
abstracts using keyword(s) &quot;"
                                    + userKeywords + "&quot;
with the URL <PRE>&quot;" + urlAddress +
"&quot;<PRE></font><BR><BR>";
                        }

                        if (errorMes == null) {
                            int pmidStart =
abstracts.indexOf(pmidTag);
                            StringBuffer tmp = abstracts;
                            html.append("<BR><br>Articles
found:<br><BR>\n");
                            StringBuffer article;
```

```
                              // 4. Formatt the articles
                              try {
                                  while (pmidStart != -1) {
                                      int endMedline =
tmp.indexOf(medlineEndTag);
                                      article = getArticle(tmp,
pmidStart, endMedline);
                                      html.append(article);


                                      tmp.delete(0, endMedline +
medlineEndTag.length());
                                      pmidStart =
tmp.indexOf(pmidTag);


                                      if (pmidStart != -1) {
                                          html.append("<HR>");
                                      }
                                  }
                              } catch (Exception e) {
                                  e.printStackTrace();
                                  errorMes = "<BR><BR><font
color=red><h1>ERROR</h1><BR>We are sorry, the system could
not retrieve the articles for PMIDs <PRE>&quot;"
                                          + sbPmids +
"&quot;<PRE></font><BR><BR>";
                                  html.append(errorMes);
                              }
                          } else {
                              html.append(errorMes);
                          }
                      } else {
                          html.append("<BR>No abstracts found!");
                      }
                  } else {
                      html.append(errorMes);
                  }
              }

              appendBuildProperty(html);
              html.append("</BODY></HTML>\n");

              // 5. Print the results
              res.setContentType("text/html");
              PrintWriter out = res.getWriter();
              out.print(html);
          }

          private void printHeader(StringBuffer html) {
              html.append("<HTML>");
              html.append("<HEAD><TITLE>" + TITLE +
"</TITLE></HEAD><BODY>\n");
```

```
            html.append("<font color=red><h1>" + TITLE +
"</h1></font>\n");
            html.append("<B>Please enter a userKeywords to
search on NCBI:</B><BR><BR>\n");
            html.append("<FORM METHOD=GET>\n");
            html.append("<INPUT TYPE=TEXT NAME=" + KEYWORDS +
"><BR><BR>\n");
            html.append("<INPUT TYPE=SUBMIT VALUE=\"Search
PubMed\"><BR>\n");
            html.append("</FORM>\n");
        }

        private StringBuffer getPmids(String spec) throws
IOException {
            BufferedReader reader = new BufferedReader(new
InputStreamReader(new
URL(spec).openConnection().getInputStream()));
            StringBuffer sbPmids = new StringBuffer();
            String pmid;
            String s = null;

            while ((s = reader.readLine()) != null) {
                if (pmidRE.match(s)) {
                    pmid = pmidRE.getParen(1);
                    sbPmids.append(pmid + ",");
                }
            }
            reader.close();
            final int length = sbPmids.length();

            if (length > 0) {
                sbPmids.delete(length - 1, length);
                return sbPmids;
            } else {
                return null;
            }
        }

        private StringBuffer getAbstracts(String urlAddress)
throws IOException {
            BufferedReader citReader =
                    new BufferedReader(new
InputStreamReader(new
URL(urlAddress).openConnection().getInputStream()));
            StringBuffer absSb = new StringBuffer();

            String s;
            while ((s = citReader.readLine()) != null) {
                absSb.append(s);
            }
            return absSb;
        }
```

```
      private StringBuffer getArticle(StringBuffer tmp, int
pmidStart, int endMedline) {
            StringBuffer articleTmp = new StringBuffer();
            String pmid1 = tmp.substring(pmidStart +
pmidTag.length(), tmp.indexOf(pmidEndTag));

            int titleStart = tmp.indexOf(titleStartTag);
            int titleEnd = tmp.indexOf(titleEndTag);

            if (titleEnd < 0 || titleEnd > endMedline)
                titleEnd = tmp.indexOf(abstractTag);

            int abstractStart = tmp.indexOf(abstractTag);

            if (titleEnd < 0 || titleEnd > endMedline) {
                titleEnd = tmp.indexOf(firstAuthorStartTag);
            }

            if (titleEnd < 0 || titleEnd > endMedline) {
                titleEnd = tmp.indexOf(firstAuthorStartTag);
            }
            String title = null;

            if (0 <= titleStart && titleStart < endMedline) {
                titleStart += titleStartTag.length();
                title = tmp.substring(titleStart,
titleEnd).replaceAll("(\\s+)", " ");
            }

            int end = tmp.indexOf(abstractEndTag);
            String tmpAbstractTag = abstractEndTag;

            if (end < 0 || end > endMedline) {
                end = tmp.indexOf(firstAuthorStartTag);
                tmpAbstractTag = firstAuthorStartTag;

                if (end < 0 || end > endMedline) {
                    end = tmp.indexOf(authorEndTag);
                    tmpAbstractTag = authorStartTag;
                }
            }

            String article = null;
            if (abstractStart + tmpAbstractTag.length() <= end)
{
                article = tmp.substring(abstractStart +
tmpAbstractTag.length(), end).replaceAll("(\\s+)", " ");
            }

            int authorStart = tmp.indexOf(authorStartTag);
            String authors = null;

            if (0 <= authorStart && authorStart < endMedline) {
```

```
                authorStart += authorStartTag.length();
                int authorEnd = tmp.indexOf(authorEndTag);
                authors = tmp.substring(authorStart,
authorEnd).replaceAll(authorStartTag, ",
").replaceAll(firstAuthorStartTag, ", ");
            }

            int srcStart = tmp.indexOf(srcTag);
            String journal = null;
            if (0 <= srcStart && srcStart < endMedline) {
                journal = tmp.substring(srcStart +
srcTag.length(), endMedline);
            }

            // Create the output
            articleTmp.append("<a href=\"" + PUBMED_ARTICLE_LK
+ pmid1 + "\">" + pmid1 + "</a>").append("<BR>");
            articleTmp.append("<U>Journal</u>: ");
            articleTmp.append(journal != null ? journal : "No
journal listed").append("<BR>");
            articleTmp.append("<u>Authors</u>: ");
            articleTmp.append(authors != null ? authors : "No
authors listed").append("<BR>");
            articleTmp.append("<u>Title</u>: ");
            articleTmp.append(title != null ? highlight(title,
params) : "No title").append("<BR>");
            articleTmp.append("<u>Abstract</u>: ");
            articleTmp.append(article != null ?
highlight(article, params) : "No article").append("<BR>");
            return articleTmp;
        }

        private String highlight(String articleText, String[]
terms) {
            String highlightedText = new String(articleText);
            for (int i = 0; i < terms.length; i++) {
                final String term = terms[i];
                final char[] chars = term.toCharArray();

    // Create the regular expression to find search terms
    // irrespective of case

                StringBuffer sb = new StringBuffer("(");
                for (char c : chars) {
                  char charUp = Character.toUpperCase(c);
                  char charLo = Character.toLowerCase(c);

sb.append("(").append(charLo).append("|").append(charUp).appe
nd(")");
                }
                sb.append(")");
                final String regex = "^" + sb.toString() +
"|[^a-zA-Z]" + sb.toString();
```

```
                // Replace the text by a HTML FONT tag
   // that wraps the term found
                RE re = new RE(regex);
                highlightedText = re.subst(highlightedText,
"\\\\<b><font style=\"\\\\+2\" color=\"" + COLOR[i] +
"\">$0</font></b>",
                        RE.REPLACE_BACKREFERENCES);
            }
            return highlightedText;
        }
```

In this Chapter, we have attempted to demonstrate how web applications can be created using the J2EE JSP and servlets technology based on a literature search and retrieval service that is indispensable for today's fast paced scientific research environment. In particular, we created a web application that provides the same powerful search capabilities provided by the NCBI PubMed server but further enhanced it by displaying the abstracts for each of the matching articles right up front and highlighting the search terms in the abstract. The rationale behind this strategy was that researchers may find it difficult to recognize the relevance of an article to their area of research simply by looking at the article title. If the abstract was displayed and the search terms were highlight and color coded, it becomes much easier to understand the context in which the abstract is relevant vis-à-vis the input search terms. This design saves the researcher a few extra clicks and makes data more readable and useful.

Note: This Chapter uses resources referred to in the Appendix: Setting up Apache ant and Apache Tomcat.

Summary

The ability to query and mine the rich scientists datasets in PubMed is a powerful way to further experimental science using a hypothesis driven research methodology where researchers build on scientific findings reported by scores of researchers around the world. In this Chapter, we have demonstrated how to create a web application with Java *Servlet/JSP* technology to access PubMed data and how to enhance the functionality provided by the resource. Processing and presentation of biomedical data in ways that provide additional benefit for the researcher is a fundamental contribution of information technologies and it is hoped that this Chapter has illustrated a small example of how this can be accomplished.

Questions and Exercises

1. Visit the NCBI PubMed website and become familiar with the service. Try out searches with different keywords and view the results using the various available Display (Brief, Abstract, Citation, XML, etc.), Sort by (Pub Date, First Author, Last Author, etc.) and Limits (Dates, Type of Article, etc.) options. Think of ways you can enhance the capabilities of the service from the user's point-of-view.

2. PubMed abstracts are a powerful source of data on protein-protein interaction networks. For example, two or more proteins mentioned in the same sentence within an abstract most likely interact with or are related to one another in some fashion. Enhance the PubMed web application we created in the Chapter by:
   a. highlighting gene/protein names mentioned in the abstract
   b. hyperlinking protein names to an appropriate annotation resource or database on the web

One such solution can be based on the use of gene symbols defined by the HUGO Gene Nomenclature Committee (HGNC). According to HGNC convention, human gene symbols are designated by upper-case Latin letters or by a combination of upper-case letters and Arabic numerals, with some exceptions. For example, the Approved Gene Symbol for the breast cancer 1, early onset gene is BRCA1.

For the second part of the exercise, the NCBI Entrez Gene resource can be used as an annotation resource. The link to the BRCA1 gene on Entrez Gene, for example, is identified by the following URL:

http://www.ncbi.nlm.nih.gov/entrez/query.fcgi?db=gene&cmd=Retrieve&dopt=full_report&list_uids=672

3. Enhance the user interface of the web application to include the capability to:
   a. save selected abstracts on your local machine
   b. filter articles by special criteria, for example, limit journals by name (Science, Nature, etc.)

Additional Resources

- The Apache Software Foundation - http://tomcat.apache.org
- The Apache Jakarta Project - http://jakarta.apache.org/regexp/
- The Apache Ant Project - http://ant.apache.org/
- Entrez - http://www.ncbi.nlm.nih.gov/Database/index.html
- HUGO Gene Nomenclature Committee - http://www.gene.ucl.ac.uk/nomenclature/
- Java Servlet API Specification 2.2 - http://java.sun.com/products/servlet/download.html
- JavaServer Pages[tm] Technology - White Paper - http://java.sun.com/products/jsp/whitepaper.html
- The Java Servlet API White Paper - http://java.sun.com/products/servlet/whitepaper.html
- Java Servlet Technology - http://java.sun.com/products/servlet/index.jsp
- PubMed Help website - http://www.ncbi.nlm.nih.gov/books/bv.fcgi?rid=helppubmed.chapter.pubmedhelp
- RFC 2616 - http://www.w3.org/Protocols/rfc2616/rfc2616.html
- RFC 3875 - http://www.rfc-archive.org/getrfc.php?rfc=3875

Selected Reading

The HUGO Gene Nomenclature Database, 2006 updates. Eyre TA, Ducluzeau F, Sneddon TP, Povey S, Bruford EA and Lush MJ. Nucleic Acids Res. 2006 Jan 1;34(Database issue):D319-21.

Guidelines for human gene nomenclature (1997). HUGO Nomenclature Committee. White JA, McAlpine PJ, Antonarakis S, Cann H, Eppig JT, Frazer K, Frezal J, Lancet D, Nahmias J, Pearson P, Peters J, Scott A, Scott H, Spurr N, Talbot C Jr, Povey S. Genomics. 1997 Oct 15;45(2):468-71.

Chapter V

Creating a Gene Prediction and BLAST Analysis Pipeline

Introduction

Gene prediction and *gene annotation* are fundamental aspects of genome-sequencing projects and discovery research. These activities involve determination of complete *gene structures* from the raw DNA sequence and attributing functions to them, by way of computational methods, at least as a first step. These methods try to implement an understanding of the way in which the structural elements such as *coding*, *non-coding* and *regulatory elements* are organized within genes, to extract meaningful information from raw nucleotide sequences.

Gene prediction programs, specifically, are designed to recognize genetic signals that are embedded in DNA sequences to make predictions about gene structure. We will explore gene prediction programs in more detail in this Chapter and build an analytic pipeline that will tie gene prediction and the BLAST application we built in earlier Chapters.

Gene Prediction Programs

Gene prediction methods that rely only on information that is encoded in the sequence itself to make predictions are called *ab initio* (Latin: from the beginning) methods. These methods use signals within DNA such as *splice sites*, *start* and *stop codons*, *promoters* and *terminators* of *transcription*, *polyadenylation sites*, *ribosomal binding sites*, *CpG islands*, and various *transcription factor binding sites* to predict the presence of *exons*. ab initio methods such as *Genscan* rely on probabilistic models known as *Hidden Markov models* (HMMs) to discern patterns within DNA. An HMM models the different states that a DNA sequence can exist in and the transition probabilities between the states. The different states of DNA are the ones enumerated above such as promoter, *intron*, *exon* etc. The term 'Hidden' comes from the fact that the sequence itself is visible but the states are hidden.

DNA Transcription and Translation

Although a detailed treatment of these subjects are out of the scope of this book, an introduction of the basic concepts in essential to understand the biology and behavior of the DNA and RNA. We had mentioned the terms transcription and translation in the last section. Transcription is the process by which a DNA molecule is copied into an RNA molecule, while translation is the process by which the RNA sequence is used by the cellular machinery to synthesize proteins.

Transcription may result in one of three types of RNA: Messenger RNA (mRNA), transfer RNAs (tRNA) or ribosomal RNA (rRNA). mRNA molecules serve as 'messengers' that specify the code for the synthesis of amino acids (during translation) and therefore the name messenger RNA. tRNAs form covalent attachments to individual amino acids and recognize the encoded sequences of the mRNAs to allow correct insertion of amino acids into the elongating polypeptide chain during translation. rRNAs are assembled together with numerous proteins to form complexes known as ribosomes. Ribosomes engage mRNAs and form a catalytic domain into which the tRNAs enter with their attached amino acids. The proteins of the ribosomes catalyze all of the functions of polypeptide synthesis.

During the process of transcription, the DNA double helix unwinds and one strand serves as the template for the synthesis of the RNA strand. Either strand can serve as the template - which strand becomes the template depends on a combination of transcription initiation and termination signals such as promoter and enhancer sequences that are present on the DNA. Transcription is actually a polymerization reaction in which individual nucleotides are linked together by an enzymatic reaction (catalyzed by the enzyme *RNA polymerase*) into a chain to form RNA.

In nature, these processes are orchestrated in a finely tuned and regulated manner involving an intricate interplay of a large number of proteins, which recognize specific signals and patterns on the sequences they bind. An example is what are known as *CpG islands*, which are regions within DNA that often occur near the beginning of genes, where the frequency of the *dinucleotide* CG (that is, the nucleotide bases *cytosine* and *guanine*) is more than in the rest of the genome

We had also mentioned exons and introns and these are simply terms used to refer to regions of DNA that code for or don't code for proteins respectively. To elaborate, higher organisms (eukaryotes) have what are called "*split genes*", that is, a large proportion of their genes are not continuous linear entities, but instead may be interrupted throughout their length by sequences that do not code for protein. A piece of DNA may therefore contain coding sequences with intervening non-coding sequences. The intervening non-coding segments are called the introns and do not code for protein. The coding sequences are exons and do code for protein. For example, the Cystic Fibrosis transmembrane regulator (CFTR) gene's coding regions (exons) are scattered over 250,000 base pairs of genomic DNA and is made up of 27 exons. During transcription, introns are removed from the CFTR gene and exons are pieced together by a process known as *RNA splicing* to form a 6100-bp mRNA transcript that is translated into the 1480 amino acid sequence (the CFTR protein). In contrast, the 384 nucleotide human pancreatic ribonuclease gene is intronless and codes for a 128 amino acid protein. A highly schematic view of the RNA splicing process is show in **Fig. 5.1**.

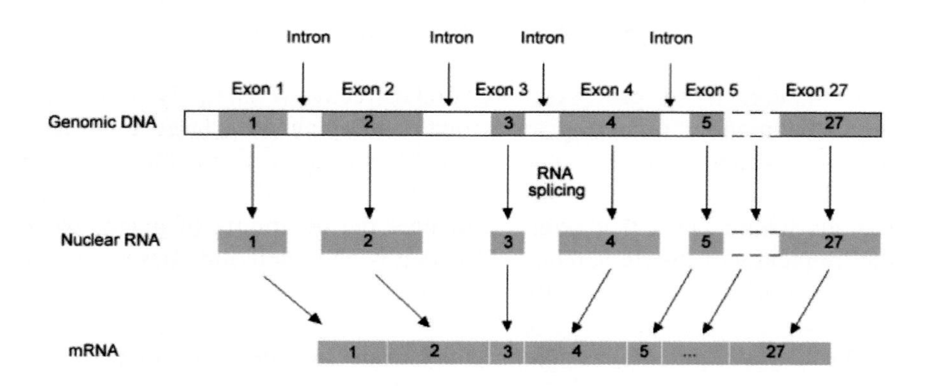

Fig. 5.1: Schematic of RNA splicing

Gene Prediction with Genscan

Genscan is one of the most effective among the many exon prediction programs to date. In this Chapter, we will build an application that will allow users to perform Genscan-based predictions on an unknown piece of DNA and analyze the predicted genes and peptides with BLAST using the SwingBlast application that we wrote earlier. The rationale to combine the two programs into this pipeline is simple – once we know that a newly sequenced stretch of DNA probably contains potential coding regions, we would like to know what peptides they may code for and what functions they perform. As we learned in Chapter 2, a BLASTX analysis of a nucleotide sequence, for example, compares a nucleotide query sequence translated in all reading frames against a protein sequence database and produces matches to known proteins. This information in turn provides clues to the probable function of an unknown peptide sequence. The integrated Genscan and BLAST pipeline can be used to perform such functional characterization of newly sequenced DNA fragments.

Genscan was written by Chris Burge and Samuel Karlin at the Department of Mathematics, Stanford University. Genscan utilizes the same basic signals described earlier to build complete *gene structures* (that is, introns + exons) from human genomic sequences. Specifically, these include transcriptional, translational and splicing signals (including elements present in most eukaryotic promoters such as the *TATA box* and cap site), as well as length distributions and compositional features of exons, introns and *intergenic* regions. Importantly, Genscan also makes use of the many substantial differences in gene density and structure based on GC composition of the human genome. For example, it is known that gene density in GC rich regions is five times higher than in regions with moderate GC content and ten times higher in rich AT rich regions. Four categories of DNA were identified based on their GC content:

1. < 43% GC
2. 43 -51% GC
3. 51 - 57% GC
4. > 57% GC

These are known as *isochores*. Thus, if the input genomic sequence has a GC content of 45 % it is said to have an isochore value of 2. ab initio programs traditionally have been poor at predicting genes in regions containing multiple genes, especially when present on both DNA strands. Genscan addresses these problems by using an explicitly double-stranded genomic sequence model, which has the likelihood of genes occurring on both DNA strands. Second, while most programs assume the presence of exactly one complete gene in the input sequence, Genscan treats the more general case in which the sequence may contain a partial gene, a complete gene, multiple complete (or partial) genes on either strand, or no gene at all. A significant difference in Genscan also is the incorporation of splice donor signal information based on the mechanism of donor splice site recognition in pre-mRNA sequences by *U1 small nuclear ribonucleoprotein particle* (U1 snRNP).

Running Genscan Analyses

Running and interpreting a Genscan analysis is rather straightforward. Point your browser to the Genscan server at MIT: http://genes.mit.edu/GENSCAN.html (**Fig. 5.2**). For this exercise we will use a 175 kilobase human bacterial artificial chromosome (BAC) with the accession number AC092818 from NCBI. Genscan has been 'trained' to work with vertebrate, arabidopsis and maize sequences (**Fig. 5.3**). Since we are analyzing a human BAC, we choose the vertebrate option. We will use the default sub-optimal exon cut-off value of 1 for our purposes. This value defines the threshold, which determines if exons that do not meet the criteria (sub-optimal exons) will be shown or not.

You can give a sequence name if you are analyzing a large number of sequences and want to label each output by a unique identifier. In this case, we will just use the BAC accession number (**Fig. 5.4**). The program gives an option to print out the predicted proteins or the predicted proteins along with their nucleotide sequences. We will choose the latter option (**Fig. 5.5**).

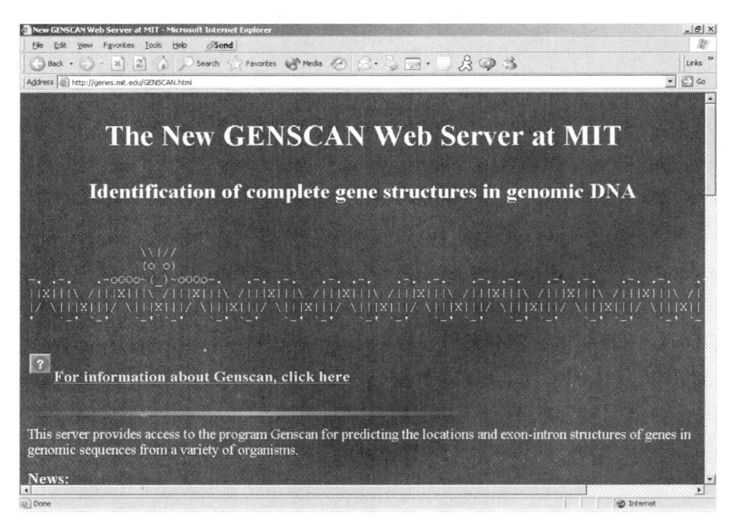

Fig. 5.2. The Genscan web server

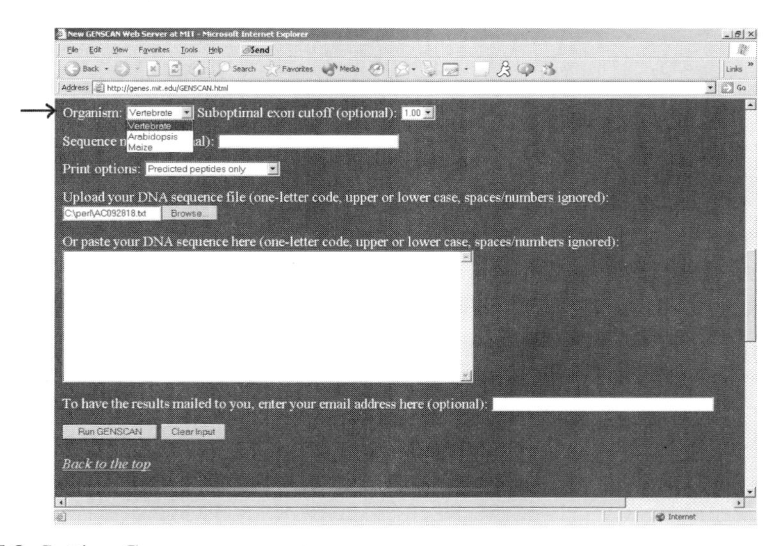

Fig. 5.3. Setting Genscan parameters

The sequence can be either uploaded or pasted directly in the text box. Uploading a sequence is more convenient if you are handling very large sequences, as is the case here (**Fig. 5.5**). Finally, you can specify an email address if you want to receive the results via email. We will hit the "Run Genscan" button and just wait to see the results in the browser. **Fig. 5.6** and **Fig. 5.7** show the results of the Genscan analysis.

Analyzing GenScan Output

The GenScan header gives information on the input sequence and the parameters used such as name, size and isochore classification (categorization based on GC content) of the sequence, and the matrix used for the analysis (HumanIso.smat). The body of the analysis consists of the predicted peptide and the corresponding CDS sequences. As is evident from the output there were eight predicted peptides in this sequence. The complete gene structure of each peptide is listed after the header (**Table 5.1**).

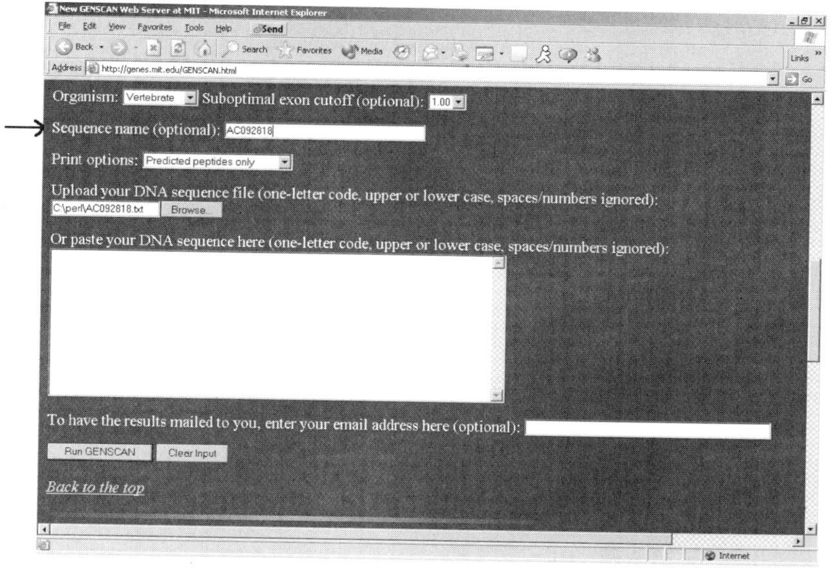

Fig. 5.4. Entering an identifier

Table 5.1. Gene structures

GENSCAN 1.0 Date run: 16-May-105 Time: 21:52:50

Sequence gi : 175100 bp : 40.28% C+G : Isochore 1 (0 - 43 C+G%)

Parameter matrix: HumanIso.smat

Predicted genes/exons:

Gn.Ex	Type	S	.Begin	...End	.Len	Fr	Ph	I/Ac	Do/T	CodRg	P....	Tscr..
1.01	Init	+	3609	3682	74	2	2	113	45	48	0.319	3.59
1.02	Intr	+	3826	3904	79	1	1	100	43	33	0.019	-1.37
1.03	Intr	+	9758	9904	147	1	0	134	35	101	0.071	8.91
1.04	Intr	+	10302	10435	134	2	2	4	75	63	0.032	-4.88
1.05	Intr	+	12763	12979	217	1	1	97	84	78	0.265	5.88
1.06	Intr	+	15363	15421	59	2	2	95	46	106	0.089	3.86
1.07	Intr	+	18293	18483	191	2	2	39	56	127	0.037	2.91
1.08	Term	+	26161	26237	77	2	2	57	43	105	0.020	-0.08
1.09	PlyA	+	27474	27479	6							1.05
2.03	PlyA	-	27633	27628	6							1.05
2.02	Term	-	48266	47967	300	2	0	-7	36	432	0.957	22.74
2.01	Init	-	49500	49009	492	0	0	64	55	181	0.650	7.60
2.00	Prom	-	50548	50509	40							-4.85
3.00	Prom	+	52752	52791	40							-5.65
3.01	Init	+	54566	54649	84	1	0	74	82	65	0.451	5.37
3.02	Intr	+	59721	59785	65	2	2	69	78	42	0.152	-2.10
3.03	Intr	+	67507	67704	198	1	0	76	46	169	0.934	9.04
3.04	Intr	+	68259	68338	80	0	2	91	75	20	0.892	-0.72
3.05	Intr	+	68461	68595	135	2	0	101	89	55	0.893	6.52
3.06	Term	+	73137	73264	128	1	2	103	39	67	0.328	0.76
3.07	PlyA	+	73438	73443	6							1.05

The most important aspects f this table are the gene and exon number, the type of exon, the strand information (+/-), the background and end positions, the length of each exon in basepairs, the frame and the scores. The key to the abbreviations is provided at the end of the output (**Table 5.2**).

Table 5.2. Abbreviations and explanations

Gn.Ex	gene number, exon number (for reference)
Type	Init = Initial exon (ATG to 5' splice site) Intr = Internal exon (3' splice site to 5' splice site) Term = Terminal exon (3' splice site to stop codon) Sngl = Single-exon gene (ATG to stop) Prom = Promoter (TATA box / initation site) PlyA = poly-A signal (consensus: AATAAA)
S	DNA strand (+ = input strand; - = opposite strand)
Begin	beginning of exon or signal (numbered on input strand)
End	end point of exon or signal (numbered on input strand)
Len	length of exon or signal (bp)
Fr	reading frame (a forward strand codon ending at x has frame x mod 3)
Ph	net phase of exon (exon length modulo 3)
I/Ac	initiation signal or 3' splice site score (tenth bit units)
Do/T	5' splice site or termination signal score (tenth bit units)
CodRg	coding region score (tenth bit units)
P	probability of exon (sum over all parses containing exon)
Tscr	exon score (depends on length, I/Ac, Do/T and CodRg scores)

Each pair of peptide and CDSs (as shown below for the first set) are in Fasta format and have unique identifiers where the sequences are numbered sequentially.

```
   >gi|GENSCAN_predicted_peptide_1|325_aa
   MALISFTSPFNFIGKKSWQCITEAGFDKVDETIIFVISQSSRNVIVGEFLQDPCQGLPL
L
   KDLSSKQAANLFPWQRMEAVACDILLIMQPGHGQPAFLQGMSSRLSGAAEQVGSWSMRS
Q
   RHSLLWSVPEPVQQAGFLFPEALQSAGCFLPSNIGLQVLQFWTLGLTSVVCQGLSGLWP
Q
   IEGCTVGFSTFEVLGLGLASLLLSLQTAYCGTSPCDHSSSLSDSKAAVLENIGLLPLTH
L
   SECSRGGTQTGISGLKTELGAKVARVCQAEYGGESHAEREFWTPTEESLRVYKRGLISS
A
   SGISVDHGSLPEGLTKTFIPEGYEP

>gi|GENSCAN_predicted_CDS_1|978_bp
atggccctaatcagttttacatctccgtttaattttattggaaagaagagctggcaatgc
atcacagaggccggctttgacaaagtggatgaaacaattatcttcgttatcagccaaagc
agtagaaatgtgatagttggggaattttgcaggacccatgccagggcttacctctgcta
aaggatttgtcctcaaagcaggcagcaaatctgttcccttggcagaggatggaagccgtg
gcttgtgacattctcctgataatgcagccaggccacgggcagccagcatttctgcagggg
atgagctccaggctcagtggggcagcagagcaagtggggagctggtccatgaggagtcag
cgtcattccttgctgtggtctgttcctgaaccagtccaacaggctggcttcctgttccca
gaagccctccaaagtgctggatgcttcctgccatcgaacattggactccaagttcttcag
ttttggactcttggacttacatcagtggtttgccaggactctcaggcctttggcctcag
```

```
attgaaggctgcactgtcggcttctctacttttgaggttttgggactcggactggcttcc
ttgctcctcagcttgcagacagcctattgtgggacttcaccttgtgatcattccagcagc
ctttcggattccaaagcggctgtcctggaaaatatagggctccttccactaacccacctc
tctgaatgcagcagaggtggaacccagacagggatcagtgggttaaagacagagctggga
gccaaggtagccagagtttgccaggcagagtatggcggagagagccacgcagagagagaa
ttctggacacctacggaggaatctcttcgagtatataaaagaggactgatcagcagtgca
tcaggtatctctgttgatcatggttctttacccgaaggactgactaaacctttattcct
gaagggtatgaaccatag
```

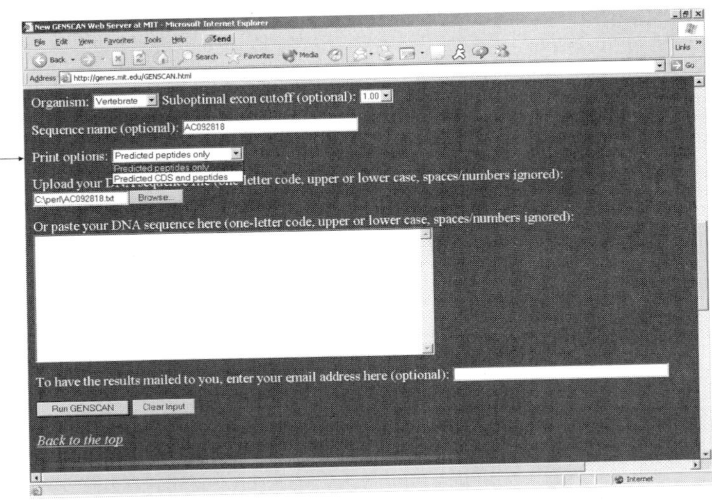

Fig. 5.5. Printing peptides and the corresponding coding sequences (CDS)

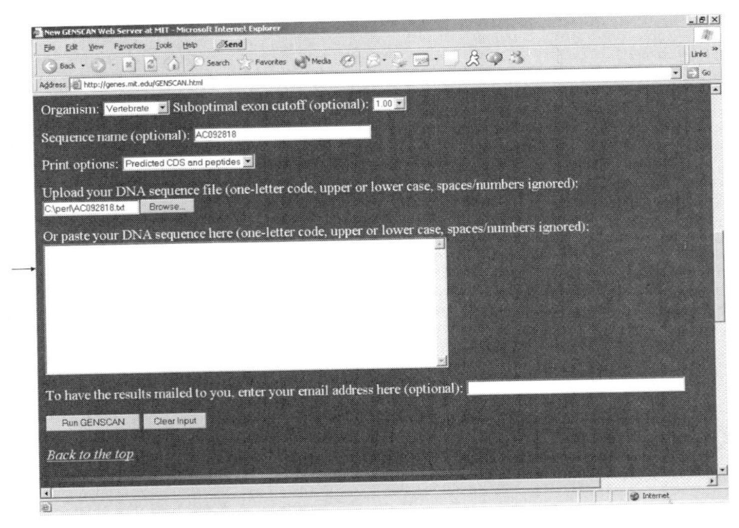

Fig. 5.6. Uploading the BAC sequence

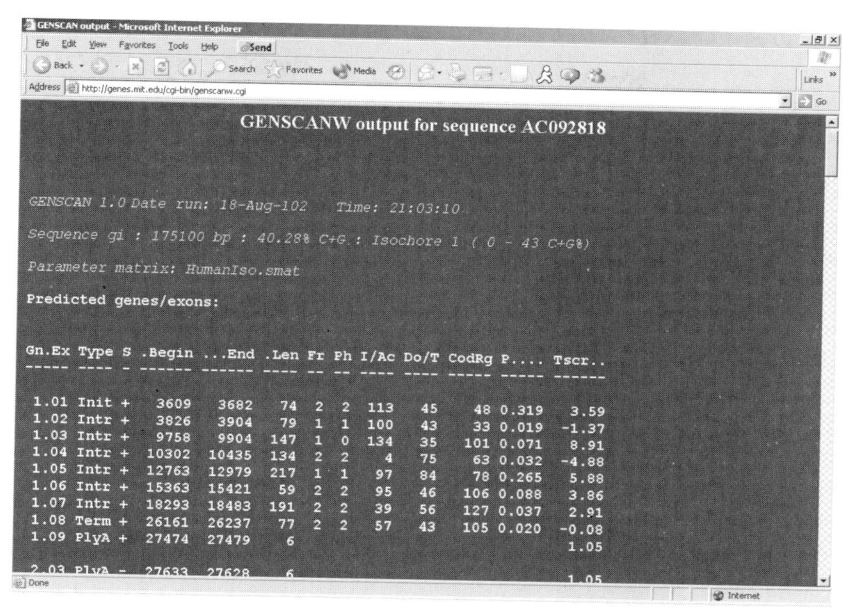

Fig. 5.7. Genscan output: Header information

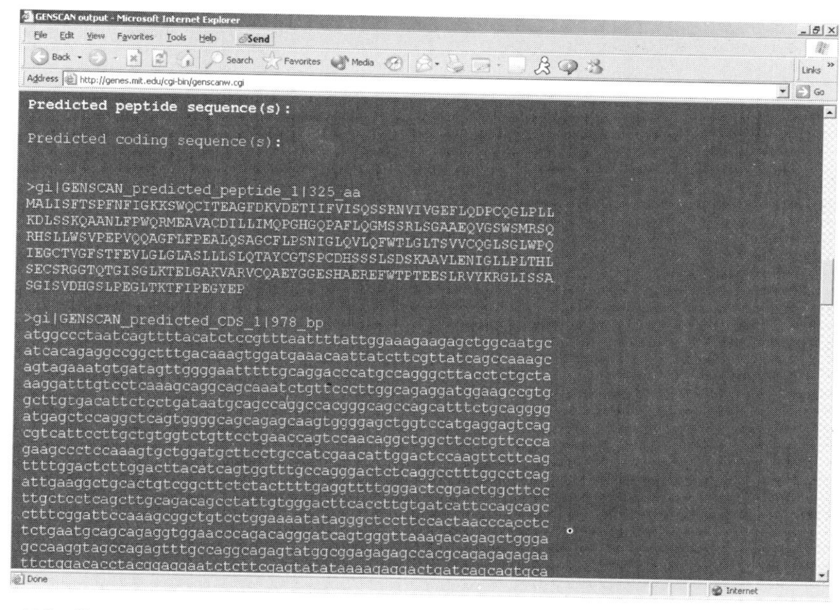

Fig. 5.8. Genscan output: predicted sequences

Creating SwingGenscan

The `SwingGenScan` application is composed of four packages as described below:

- `org.jfb.genscan`: contains the Genscan API that provides a framework for a Genscan implementation. It makes the implementation more flexible by allowing us to optimize, thread, or queue requests and perform other manipulations without having to change the whole application; the way the implementation works is transparent to the application.
- `org.jfb.jgenscan`: a Genscan implementation of the framework defined by the `org.jfb.genscan` package.
- `org.jfb.util`: contains classes for performing operations such as extracting the peptide and genes from a Genscan prediction.
- `org.jfb.swinggenscan`: contains all the classes to build the `SwingGenScan` application.
- `GenScanResult`: contains the parsed peptide and the gene predictions.
- `ResultDialog`: a JDialog window that displays the result of Genscan operation. In this window, users can select one or more sequences to place into the BLAST pipeline using `SwingBlast`.
- `SwingGenScan`: the main application window where users can select the parameters for running a Genscan prediction against a chosen nucleotide sequence

The goal of this Chapter is to create a gene prediction and annotation pipeline which enables a user to perform gene prediction followed by further downstream analysis of the predicted gene and peptide sequences using BLAST. SwingGenScan uses SwingBlast to send Genscan predicted sequences for BLAST analysis. To enable this, we have modified `SwingBlast` version 2.5 that we created in Chapter 3 and separated the functionality provided by that application into four packages that we will use in `SwingGenScan`:

- `org.jfb.blast`: provides the BLAST API
- `org.jfb.jqblast`: provides an implementation of the BLAST API
- `org.jfb.util`: contains classes that provide functions that can be shared by more than one application (to enable future code reuse). For instance, the class QueryHelper in this package contains two methods

(sendQuery and postQuery) to send GET or POST HTTP requests and the HTML result back.

- `org.jfb.swingblast3`: is the new refactored SwingBlast application. Since this is a major change, we have named this version 3.

The four classes can be packaged into a jar file called `swingblast.jar`. The jar file can serve as a library whose functionality can be used like any other Java library by placing it in the Java classpath. The structure of the `SwingGenScan` application is shown in **Fig. 5.9**.

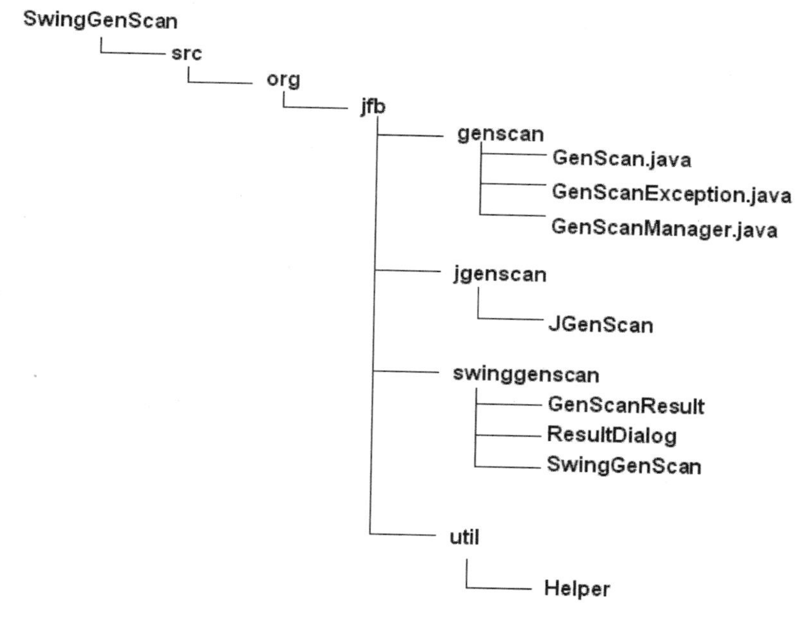

Fig. 5.9. The SwingGenScan application structure

Writing the Code for SwingGenScan

The `org.jfb.genscan` package contains the following Java classes:

`GenScan.java`

`GenScanException.java`, and

```
GenScanManager.java
```

As described earlier, this package contains the API that provides a framework for a Genscan implementation. Let's look at the code for the first Java class `GenScan` located in the file `Genscan.java` (**Listing 5.1**).

Listing 5.1. Code for Java class GenScan

```
package org.jfb.genscan;

import java.util.HashMap;
import java.util.Observable;

public abstract class GenScan extends Observable {
    public abstract Object submitQuery(Map parameters)
throws GenScanException;

    public abstract Object requestResult(Object identifier)
            throws GenScanException,
IllegalArgumentException;
}
```

When we run a Genscan analysis, we would like to know the status of the Genscan operation - has the request been submitted and if so, is the sequence currently in process, or has it encountered an error? The `Genscan` class provides a simple way of being notified of events through the use of the *observer pattern* as described in Chapter 2.

Next we define the GenScanManager class, whose purpose is to provide an instance of GenScan (**Listing 5.2**). As we'll see later, an implementation of the GenScan API will call the GenScanManager's register method to register itself as the default GenScan implementation.

Remember, we don't want to modify our code if we change the GenScan implementation to provide a multi-threaded, queued and multi-server implementation in the future. So to load our GenScan implementation we just pass the full name of the Java class to load, through the JVM system property (defined as "genscanClass.driver") using the –D option as explained earlier in Chapter 3. Another way is to call `Class.forName` ("full name of the Java class") to have the Java *classloader* locate the implementation and load it into the *JVM*. The reader will notice that the `createGenScan()` is *thread safe*, which means that a different instance of the Java `GenScan` implementation will be loaded for each thread and therefore it will not be a problem while accessing shared resources. For the same reason, multiple Genscan analyses can be run in a multi-threaded application. To return an instance of the implementation of GenScan we then use the *Java reflection API* (defined in *java.lang.reflect* package) to retrieve the constructor and create a new object of the GenScan implementation here called `JgenScan`.

Listing 5.2. GenScanManager.Java

```
package org.jfb.genscan;

public class GenScanManager {
     private static String genscanClass = null;
     private static boolean initialized = false;

     public   static   synchronized   void   register(GenScan
genscan) {
          genscanClass = genscan.getClass().getName();
          initialized = true;
     }

     private static void loadInitialDrivers() {
          final           String           driver           =
System.getProperty("genscanClass.driver");
          if (driver == null)
               return;

          try {
               System.out.println("GenScanManager.Initialize:
loading " + driver);
               Class.forName(driver);
          } catch (Exception e) {
               System.out.println("GenScanManager.Initialize:
load failed: " + e);
          }
     }

     public   static   GenScan   createGenScan()   throws
GenScanException {
          if (!initialized) {
               initialized = true;
               loadInitialDrivers();
          }
          if (genscanClass == null)
               throw new GenScanException("There is no driver
configured! "
                         + "Please use genscanClass.driver Java
property or Class.forName" +
                         " to load the driver class.");
          try {
```

```
            // In a multi thread environment we need to
//make sure that the class is loaded
            final Class aClass = (Class)
Class.forName(genscanClass, true,

Thread.currentThread().getContextClassLoader());
            return (GenScan) aClass.getConstructor(new
Class[]{}).newInstance(new Object[]{});
        } catch (Exception e) {
            throw new GenScanException(e);
        }
    }
}
```

Next, we need to be able to get an instance of `GenScan`, or more specifically, an instance of the implementation that fulfills our Java `GenScan` declaration requirements. The design of the GenScan framework provided by the API we wrote is to make the implementation transparent to the user. For example, the implementation uses an HTTP server to run the Genscan analysis and to retrieve the result. This entire process is shielded from the user. The user simply calls the `submitQuery` method with a Map of parameters and requests a result using an object identifier.

The code below loads the class for the Genscan implementation:

```
(Class aClass = (Class) Class.forName(genscanClass, true,

Thread.currentThread().getContextClassLoader());
            return (GenScan) aClass.getConstructor(new
Class[]{}).newInstance(new Object[]{});
```

We use *Java reflection* to retrieve a `Class` instance of the class defined by the name `genscanClass` by calling the static method `forName` from class `Class` and we cast it to `Class`. Then we use the `Class` instance we retrieved to construct an instance of that class by calling the `getConstructor` method that we cast also to type `GenScan`. *Casting* an object means forcing the object to be of a certain Java type. Of course, the type one wants to cast an object into must be one that the object inherits from. The new type can be an interface, an abstract class or a super class type. Casting is done in Java by putting the new type in parentheses before the object as shown above.

Note the static method in `GenScanManager.Java`:

```
public static synchronized void register(GenScan genscan) {
        genscanClass = genscan.getClass().getName();
        initialized = true;
    }
```

This method allows any implementation to register itself to the `GenScanManager` by calling it with an instance of an implementation of `GenScan` in a *static statement*. The method just stores the full Java class name of the implementation of `GenScan` by using Java reflection (`getClass()` method) on an object. The name will be then used by the `createGenScan()` method to provide an instance of `GenScan`.

Finally, the `GenScanException` class handles any exceptions that may arise during the operation of `Genscan` (**Listing 5.3**).

Listing 5.3. GenScanException class

```
package org.jfb.genscan;

public class GenScanException extends Exception {
    public GenScanException() {
    }

    public GenScanException(String message) {
        super(message);
    }

    public    GenScanException(String    message,    Throwable
cause) {
        super(message, cause);
    }

    public GenScanException(Throwable cause) {
        super(cause);
    }
}
```

Next we implement `Genscan` as shown in **Listing 5.4.** In the `JGenScan` class, the `register()` method is called by `createGenScan()` in case no Java class name for any implementation has been provided. Next the method `loadInitialDrivers()` will attempt to first retrieve the full Java class name of the implementation by looking at a JVM system property passed through the *JVM* as argument using the *–D option* as explained before:

```
java –DgenscanClass.driver=org.jfb.jgenscan.JgenScan
```

The line above will define in the system the property `genscanClass.driver` with the value `org.jfb.jgenscan.JgenScan`. We get the system property back in the Java code like this:

```
    System.getProperty("genscanClass.driver");
```

If the value found is not *null*, the method will then attempt to load the class through a class method call – `Class.forName()`. If `JGenScan` is not in the Java classpath, then the Java classloader will fail to load the class and will throw a `ClassNotFoundException`. So it is important to make sure that you declare `JGenScan` in the Java classpath. The method `forName()` has the effect of initializing the class implementing `GenScan`. Part of the initialization is to run the static statements and set up the static fields or constants.

Listing 5.4. The JGenScan class

```
package org.jfb.jgenscan;

import org.jfb.genscan.GenScan;
import org.jfb.genscan.GenScanException;
import org.jfb.genscan.GenScanManager;
import org.jfb.util.QueryHelper;

import java.io.UnsupportedEncodingException;
import java.net.URLEncoder;
import java.util.ArrayList;
import java.util.Collection;
import java.util.HashMap;

public class JGenScan extends GenScan {
    private static final String GENSCAN_HOSTNAME =
"genes.mit.edu";
    private static final String GENSCAN_PATH = "/cgi-
bin/oldgenscanw.cgi";
    private static final int GENSCAN_PORT = 80;
    private static final String GENSCAN_URL = "http://" +
GENSCAN_HOSTNAME + ":" + GENSCAN_PORT + "/" + GENSCAN_PATH;

    static {
        System.out.println("Registering " +
JGenScan.class);
        GenScanManager.register(new JGenScan());
    }


    private static Map reqIdToResultFileName = new
HashMap();
    private Collection currentRunningGenScan = new
ArrayList();
    private static final int NUMBER_OF_SECOND = 3000;

    public Object submitQuery(Map parameters) throws
GenScanException {
        final String urlapiQuery =
createUrlapiQuery(parameters);
        setChanged();
        notifyObservers("Submitting the job to the server
with query\n" + urlapiQuery);
        Runnable runnable = new Runnable() {
            public void run() {
                Object res;
                try {
                    res =
QueryHelper.sendQuery(urlapiQuery, GENSCAN_URL, true);
                } catch (Throwable e) {
                    res = new GenScanException("Problem
with URL " + GENSCAN_URL, e);
```

```
                }
                final String key = "" + this.hashCode();
                synchronized (reqIdToResultFileName) {
                    System.out.println("Storing the result
...");
                    reqIdToResultFileName.put(key, res);
                }
            }
        };
        new Thread(runnable).start();
        final String key = "" + runnable.hashCode();
        currentRunningGenScan.add(key);
        return key;
    }

    public Object requestResult(Object identifier) throws
GenScanException {
        if (!currentRunningGenScan.contains(identifier))
            throw new IllegalArgumentException(identifier +
" has no corresponding result!");
        Map tmp = null;
        boolean hasFinished = false;
        int ct = 0;
        synchronized (this) {
            while (!hasFinished) {
                tmp = new HashMap(reqIdToResultFileName);
                hasFinished = tmp.containsKey(identifier);
                if (hasFinished) {

reqIdToResultFileName.remove(identifier);
                    break;
                }
                setChanged();
                notifyObservers("Waiting " +
NUMBER_OF_SECOND
                        + " seconds before re-trying (total
waiting time: "
                        + (ct += NUMBER_OF_SECOND) +
"s).");
                try {
                    wait(NUMBER_OF_SECOND);
                } catch (InterruptedException e) {
                    e.printStackTrace();
                }
            }
        }
        final Object o = tmp.get(identifier);
        if (o == null) return null;
        if (o instanceof Throwable)
            throw new GenScanException("Embedded
exception", (Throwable) o);
        return o;
    }
```

```
        private String createUrlapiQuery(Map parameters) {
            StringBuffer query = new StringBuffer();
            try {
                final Object org = parameters.get("organism");
                final Object nam = parameters.get("name");
                final              Object              sub          =
parameters.get("subOptExonCutoff");
                final              Object              dis          =
parameters.get("displayOption");
                query.append("-
s=").append(URLEncoder.encode((String)
parameters.get("sequence"), "UTF-8"));
                if (org != null) {
                    query.append("&-
o=").append(URLEncoder.encode((String) org, "UTF-8"));
                }
                if (nam != null) {
                    query.append("&-
n=").append(URLEncoder.encode((String) nam, "UTF-8"));
                }
                if (sub != null) {
                    query.append("&-
e=").append(URLEncoder.encode((String) sub, "UTF-8"));
                }
                if (dis != null) {
                    query.append("&-
p=").append(URLEncoder.encode((String) dis, "UTF-8"));
                }
            } catch (UnsupportedEncodingException uee) {
                uee.printStackTrace();
            }
            return query.toString();
        }
    }
```

Note the following piece of code in **Listing 5.4**:

```
new Thread(runnable).start();
final String key = "" + runnable.hashCode();
currentRunningGenScan.add(key);
return key;
```

Here, we are threading the process to be able to run more than one query without having to wait for the first one to finish. Also because we're running in *a* multi-threaded environment we want to synchronize the Map called `reqIdToResultFileName`, to safely save the right key with the right result and to avoid more than one thread to modify the Map at the same time that could potentially populate the Map with wrong data.

After we have submitted the query, we retrieve the result by calling the `requestResult()` method. That method will return only when the result is available. One has to make sure that a call to that method is not executed in the event-dispatching thread, because that will block the repaint of the application.

The method `requestResult()` described in **Listing 5.4** first checks that the request identifier is a valid argument. If invalid, the method will throw an exception that would allow us to track down multiple calls to the method with the same argument that could probably imply an infinite loop. We are protecting multiple threads from accessing the same block when we are checking if the request is ready, by surrounding the block with a *synchronized ()* block. The synchronization is on the current object "*this*" calling that method. That means that the JVM will set a lock (a unique token) on the current object to the thread that first entered the block. Then, until the thread inside that block releases the lock, any other threads waiting to run that piece of code will have to wait for the lock to be released. The actual processes are transparent to the developer because of the use of the *synchronized* Java keyword.

The result of the `Genscan` operation is stored in the `GenScanResult` object. This is essentially the predicted peptide and gene sequences and any additional data about the search that the user may wish to save such as the name of the server, the Genscan parameters used for the prediction as well as the time taken to execute the prediction etc. The code for the `GenScanResult` class is shown in **Listing 5.5**.

Listing 5.5. GenScanResult.Java

```
package org.jfb.swinggenscan;

public class GenScanResult {
    private String[] peptideGene = null;

    public void setPeptideAndGene(String[] pepGene) {
        peptideGene = pepGene;
    }

    public String[] getPeptideGene() {
        return peptideGene;
    }
}
```

Next, the `ResultDialog class` takes a `GenScanResult` object and displays its content.

```
    public void showResult(GenScanResult result) {
        String[] pepGene = result.getPeptideGene();
        if (pepGene == null) {
            list.setCellRenderer(new
DefaultListCellRenderer());
            list.setListData(new    String[]{"No    Results
Found"});
        } else {
            list.setCellRenderer(new MyListCellRenderer());
            list.setListData(pepGene);
        }
    }
```

The `ResultDialog class` also allows the user to run additional analyses to be run on the predicted gene and peptide sequences. In this case, we will add a functionality to perform a BLAST search on user selected Genscan predictions. To do that, we add a check box against each predicted sequence and a button called "`Run Blast`" at the bottom. Once the user selects a sequence and hits the "`Run BLAST`" button, the `SwingBlast` application we created earlier is invoked with the selected sequences in the text area of the `SwingBlast` application.

```
        runBlastButton = new JButton("Run Blast");
        runBlastButton.addActionListener(new
ActionListener() {
            public void actionPerformed(ActionEvent e) {
                if (!list.isSelectionEmpty()) {

SwingBlast3.launch(list.getSelectedValues()[0].toString());
```

```
                    }
```

The code for the `ResultDialog` class is shown in **Listing 5.6**.

Listing 5.6. ResultDialog.java

```
package org.jfb.swinggenscan;

import org.jfb.swingblast3.SwingBlast3;
import javax.swing.*;
import javax.swing.event.ListSelectionEvent;
import javax.swing.event.ListSelectionListener;
import java.awt.*;
import java.awt.event.ActionEvent;
import java.awt.event.ActionListener;

public class ResultDialog extends JDialog {
    private static final Dimension BD_PREF_SIZE = new
Dimension(530, 460);
    private JList list;
    private JButton runBlastButton;

    public ResultDialog(Frame owner) throws
HeadlessException {
        super(owner);
        setTitle("GenScan Result Dialog");
    }

    public void init() {
        list = new JList();

list.setSelectionMode(ListSelectionModel.SINGLE_SELECTION);
        list.addListSelectionListener(new
ListSelectionListener() {
            public void valueChanged(ListSelectionEvent e)
{
                if (!e.getValueIsAdjusting()) {

runBlastButton.setEnabled(!list.isSelectionEmpty());
                }
            }
        });
        JScrollPane scrollPaneArea = new JScrollPane(list);
        scrollPaneArea.setPreferredSize(new Dimension(500,
400));
        JPanel panel = new JPanel();
        panel.setLayout(new BorderLayout());
        panel.add(scrollPaneArea, BorderLayout.NORTH);

        JPanel buttonPane = new JPanel();
        buttonPane.setLayout(new BoxLayout(buttonPane,
```

```
BoxLayout.LINE_AXIS));
          buttonPane.add(Box.createHorizontalGlue());
          buttonPane.add(Box.createRigidArea(new
Dimension(10, 0)));

          runBlastButton = new JButton("Run Blast");
          runBlastButton.addActionListener(new
ActionListener() {
              public void actionPerformed(ActionEvent e) {
                  if (!list.isSelectionEmpty()) {

SwingBlast3.launch(list.getSelectedValues()[0].toString());
                  }
              }
          });
          runBlastButton.setSize(new Dimension(80, 20));
          runBlastButton.setEnabled(false);
          buttonPane.add(runBlastButton);
          panel.add(runBlastButton, BorderLayout.SOUTH);
          getContentPane().add(panel);
          setSize(BD_PREF_SIZE);
          setVisible(true);
      }

      public void showResult(GenScanResult result) {
          String[] pepGene = result.getPeptideGene();
          if (pepGene == null) {
              list.setCellRenderer(new
DefaultListCellRenderer());
              list.setListData(new    String[]{"No    Results
Found"});
          } else {
              list.setCellRenderer(new MyListCellRenderer());
              list.setListData(pepGene);
          }
      }


      private    static    class    MyListCellRenderer    implements
ListCellRenderer {
          public Component getListCellRendererComponent(JList
list,  final  Object  value,  int  index,  boolean  isSelected,
boolean cellHasFocus) {
              JPanel jPanel = new JPanel();
              jPanel.setLayout(new BorderLayout());
              final      JTextArea      textArea      =      new
JTextArea(value.toString());
              final Font sf = textArea.getFont();
              Font  f  =  new  Font("Monospaced",  sf.getStyle(),
sf.getSize());
              textArea.setFont(f);
              textArea.setLineWrap(true);
              final JCheckBox comp = new JCheckBox();
```

```
                comp.setSelected(isSelected);
                jPanel.add(comp, BorderLayout.WEST);
                jPanel.add(textArea, BorderLayout.CENTER);
                return jPanel;
            }
        }
    }
```

The SwingGenScan User Interface

The application interface is created using swing libraries. **Listing 5.7** shows the code for the `SwingGenScan` application.

Listing 5.7. SwingGenScan user interface

```
package org.jfb.swinggenscan;


import org.jfb.genscan.GenScan;
import org.jfb.genscan.GenScanException;
import org.jfb.genscan.GenScanManager;
import org.jfb.util.Helper;

import javax.swing.*;
import javax.swing.event.DocumentEvent;
import javax.swing.event.DocumentListener;
import java.awt.*;
import java.awt.event.ActionEvent;
import java.awt.event.ActionListener;
import java.util.HashMap;
import java.util.Observable;
import java.util.Observer;

public class SwingGenScan extends JFrame {
    private static final String APP_NAME = "SwingGenScan";
    private static final String APP_VERSION = "Version
1.0";
    private static final String STATUS_LABEL = "Status: ";
    private static final String STATUS_READY = "Ready";

    private static final Dimension LABEL_PREFERRED_SIZE =
new Dimension(127, 16);
    private static final Dimension COMBO_PREFERRED_SIZE =
new Dimension(60, 25);
    private static final Dimension CP_PREF_SIZE = new
Dimension(450, 410);

    private static final String[] ORGANISMS =
            new String[]{"Vertebrate", "Arabidopsis",
```

```
"Maize"};
      private static final String[] PRINT_OPTIONS =
              new    String[]{"Predicted    peptides    only",
"Predicted CDS and peptides"};
      private         static         final         String[]
SUBOPTIMAL_EXON_CUTOFF_VALUES =
              new  String[]{"1.00",  "0.50",  "0.25",  "0.10",
"0.05", "0.02", "0.01"};

      private JComponent newContentPane;
      private JTextArea sequenceArea;
      private JScrollPane scrollPaneArea;
      private JLabel statusLabel;
      private JLabel statusText;

      private JComboBox organisms;
      private JComboBox printOptions;
      private JComboBox exonCutoffs;

      private JButton clearBtn, submitBtn;

      private JMenuItem aboutItem;
      private JMenuItem quitItem;

      public SwingGenScan() {
          super();
          seqFormInit();
      }

      private void seqFormInit() {
          setTitle(APP_NAME);
          setDefaultCloseOperation(JFrame.EXIT_ON_CLOSE);
          newContentPane = new JPanel();
          newContentPane.setOpaque(true);
          newContentPane.setLayout(new BorderLayout());

  getContentPane().add(newContentPane)
          setContentPane(newContentPane);

          // Create the menu bar
          JMenuBar menu = new JMenuBar();
          JMenu swingBlastMenu = new JMenu(APP_NAME);
          quitItem = new JMenuItem("Quit");
          swingBlastMenu.add(quitItem);
          menu.add(swingBlastMenu);

          JMenu helpMenu = new JMenu("Help");
          aboutItem = new JMenuItem("About");
          helpMenu.add(aboutItem);
          menu.add(helpMenu);
          setJMenuBar(menu);
```

```
            // Create the sequence pane
            JPanel sequencePanel = new JPanel();
            JLabel sequence = new JLabel("Sequence");
            sequenceArea = new JTextArea();
            final Font sf = sequenceArea.getFont();
            Font  f  =  new  Font("Monospaced",  sf.getStyle(),
sf.getSize());
            sequenceArea.setFont(f);
            sequenceArea.setLineWrap(true);
            scrollPaneArea = new JScrollPane(sequenceArea);
            scrollPaneArea.setPreferredSize(new  Dimension(300,
200));

            sequencePanel.setLayout(new
BoxLayout(sequencePanel, BoxLayout.LINE_AXIS));
            sequencePanel.add(sequence);
            sequencePanel.add(Box.createRigidArea(new
Dimension(10, 0)));
            sequencePanel.add(scrollPaneArea);

sequencePanel.setBorder(BorderFactory.createEmptyBorder(10,
0, 10, 0));

            statusLabel = new JLabel(STATUS_LABEL);
            statusLabel.setPreferredSize(new      Dimension(50,
30));
            statusText = new JLabel(STATUS_READY);
            JPanel statusPanel = new JPanel();

statusPanel.setBorder(BorderFactory.createEmptyBorder(0,   5,
5, 5));
            statusPanel.setLayout(new BorderLayout());
            statusPanel.add(statusLabel, BorderLayout.WEST);
            statusPanel.add(statusText, BorderLayout.CENTER);

            // Lay out the buttons from left to right

            JPanel buttonPane = new JPanel();
            submitBtn = new JButton("Submit");
            clearBtn = new JButton("Clear");

            buttonPane.setLayout(new      BoxLayout(buttonPane,
BoxLayout.LINE_AXIS));
            buttonPane.add(Box.createHorizontalGlue());
            buttonPane.add(Box.createRigidArea(new
Dimension(10, 0)));
            buttonPane.add(clearBtn);
            buttonPane.add(submitBtn);

            JPanel jPanel = new JPanel();
            jPanel.setLayout(new BorderLayout());
            jPanel.setBorder(BorderFactory.createEmptyBorder(0,
10, 10, 10));
```

```
            jPanel.add(sequencePanel, BorderLayout.NORTH);
            jPanel.add(createProgramPanel(),
BorderLayout.CENTER);
            jPanel.add(buttonPane, BorderLayout.SOUTH);

            newContentPane.add(jPanel, BorderLayout.CENTER);
            newContentPane.add(statusPanel,
BorderLayout.SOUTH);
            newContentPane.setPreferredSize(CP_PREF_SIZE);
            enableFunctions(false);
            // Display the window
            pack();
            Dimension                    screenSize            =
Toolkit.getDefaultToolkit().getScreenSize();
            setLocation((screenSize.width - CP_PREF_SIZE.width)
/ 2,
                    (screenSize.height - CP_PREF_SIZE.height) /
2);
            setVisible(true);
            addListeners();
        }

        private JPanel createProgramPanel() {
            JPanel organismPanel = new JPanel();
            JLabel organismLabel = new JLabel("Organism");

organismLabel.setPreferredSize(LABEL_PREFERRED_SIZE);
            organisms = new JComboBox(ORGANISMS);
            organisms.setMaximumSize(COMBO_PREFERRED_SIZE);
            organismPanel.setLayout(new
BoxLayout(organismPanel, BoxLayout.LINE_AXIS));
            organismPanel.add(organismLabel);
            organismPanel.add(Box.createRigidArea(new
Dimension(10, 0)));
            organismPanel.add(organisms);
            organismPanel.add(Box.createRigidArea(new
Dimension(5, 0)));
            organismPanel.add(Box.createHorizontalGlue());

            JPanel exonCutoffPanel = new JPanel();
            JLabel  exonCutoffLabel  =  new  JLabel("Suboptimal
Exon Cuttoff");

exonCutoffLabel.setPreferredSize(LABEL_PREFERRED_SIZE);
            exonCutoffs                    =                   new
JComboBox(SUBOPTIMAL_EXON_CUTOFF_VALUES);
            exonCutoffs.setMaximumSize(COMBO_PREFERRED_SIZE);
            exonCutoffPanel.setLayout(new
BoxLayout(exonCutoffPanel, BoxLayout.LINE_AXIS));
            exonCutoffPanel.add(exonCutoffLabel);
            exonCutoffPanel.add(Box.createRigidArea(new
Dimension(10, 0)));
            exonCutoffPanel.add(exonCutoffs);
```

```
            exonCutoffPanel.add(Box.createRigidArea(new
Dimension(5, 0)));
            exonCutoffPanel.add(Box.createHorizontalGlue());

            JPanel printOptionsPanel = new JPanel();
            JLabel   printOptionsLabel   =   new   JLabel("Print
Options");

printOptionsLabel.setPreferredSize(LABEL_PREFERRED_SIZE);
            printOptions = new JComboBox(PRINT_OPTIONS);
            printOptions.setMaximumSize(COMBO_PREFERRED_SIZE);
            printOptionsPanel.setLayout(new
BoxLayout(printOptionsPanel, BoxLayout.LINE_AXIS));
            printOptionsPanel.add(printOptionsLabel);
            printOptionsPanel.add(Box.createRigidArea(new
Dimension(10, 0)));
            printOptionsPanel.add(printOptions);
            printOptionsPanel.add(Box.createRigidArea(new
Dimension(5, 0)));
            printOptionsPanel.add(Box.createHorizontalGlue());

            JPanel paramPanel = new JPanel();
            paramPanel.setLayout(new      BoxLayout(paramPanel,
BoxLayout.PAGE_AXIS));

            paramPanel.add(organismPanel);
            paramPanel.add(Box.createRigidArea(new Dimension(0,
5)));
            paramPanel.add(exonCutoffPanel);
            paramPanel.add(Box.createRigidArea(new Dimension(0,
5)));
            paramPanel.add(printOptionsPanel);
            paramPanel.add(Box.createRigidArea(new Dimension(0,
5)));

            return paramPanel;
        }

        private void addListeners() {
            quitItem.addActionListener(new ActionListener() {
                public void actionPerformed(ActionEvent e) {
                    System.exit(0);
                }
            });

            aboutItem.addActionListener(new ActionListener() {
                public void actionPerformed(ActionEvent e) {

JOptionPane.showMessageDialog(org.jfb.swinggenscan.SwingGenSc
an.this, APP_NAME + " " + APP_VERSION,
                          "About     "      +      APP_NAME,
JOptionPane.INFORMATION_MESSAGE);
                }
```

```
        });

        clearBtn.addActionListener(new ActionListener() {
            public void actionPerformed(ActionEvent e) {
                sequenceArea.setText("");
                enableFunctions(false);
                statusText.setText(STATUS_READY);
            }
        });

        submitBtn.addActionListener(new ActionListener() {
            public void actionPerformed(ActionEvent e) {

                Runnable runnable = new Runnable() {
                    public void run() {
                        GenScan genScan = null;
                        try {

Class.forName("org.jfb.jgenscan.JGenScan");
                            genScan                          =
GenScanManager.createGenScan();
                        }    catch    (ClassNotFoundException
cnfe) {

                            cnfe.printStackTrace();
                        } catch (GenScanException gse) {
                            gse.printStackTrace();
                        }
                        Map param = new HashMap();
                        param.put("sequence",
sequenceArea.getText());
                        param.put("organism",
organisms.getSelectedItem());
                        param.put("subOptExonCutoff",
exonCutoffs.getSelectedItem());
                        param.put("displayOption",
printOptions.getSelectedItem());
                        Object requestIdentifier = null;
                        try {
                            requestIdentifier              =
genScan.submitQuery(param);
                        } catch (GenScanException gse) {
                            gse.printStackTrace();
                        }
                        Observer observer = new Observer()
{
                            public  void  update(Observable
o, Object arg) {

SwingGenScan.this.statusText.setText(arg.toString());
                            }
                        };
                        genScan.addObserver(observer);
                        Object text = null;
```

```
                            try {
                                text                                    =
genScan.requestResult(requestIdentifier);
                            } catch (GenScanException gse) {
                                gse.printStackTrace();
                            }
                            final    GenScanResult    result    =
Helper.extractPeptideAndGene(text.toString());
                            EventQueue.invokeLater(new
Runnable() {
                                public void run() {

statusText.setText(STATUS_READY);
                                    final          ResultDialog
resultDialog = new ResultDialog(SwingGenScan.this);
                                    resultDialog.init();

resultDialog.showResult(result);
                                }
                            });
                        }
                    };
                    new Thread(runnable).start();
                }
            });

            sequenceArea.getDocument().addDocumentListener(new
DocumentListener() {
                public void insertUpdate(DocumentEvent e) {

enableFunctions(sequenceArea.getText().trim().length() > 0);
                }

                public void removeUpdate(DocumentEvent e) {

enableFunctions(sequenceArea.getText().trim().length() > 0);
                }

                public void changedUpdate(DocumentEvent e) {
                }
            });
        }


        private void enableFunctions(boolean enabled) {
            organisms.setEnabled(enabled);
            exonCutoffs.setEnabled(enabled);
            printOptions.setEnabled(enabled);
        }

        public static void main(String[] args) {
            SwingUtilities.invokeLater(new Runnable() {
                public void run() {
```

```
                        new SwingGenScan();
                    }
                });
            }
        }
```

The `SwingGenScan` user interface is shown in **Fig. 5.10**.

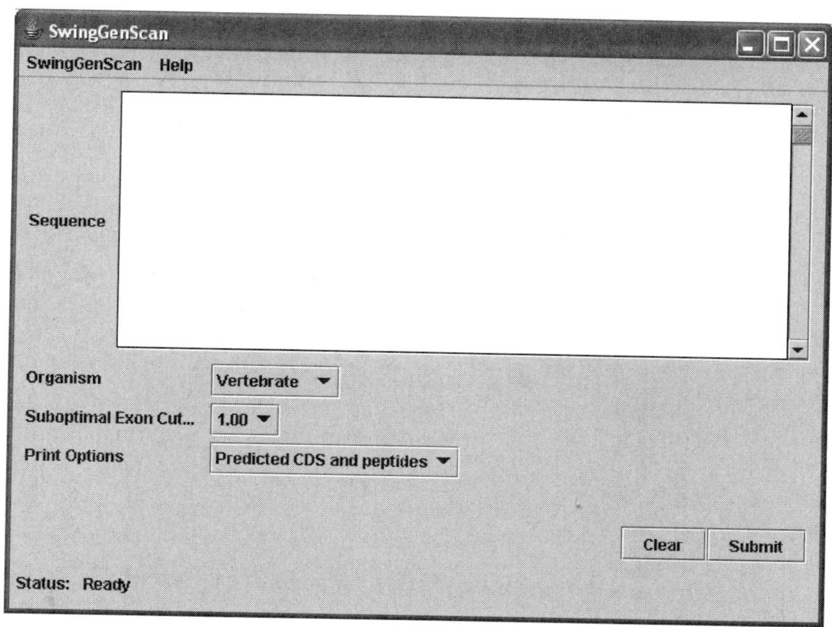

Fig. 5.10. SwingGenScan user interface

After the Genscan prediction has finished, we need to parse the raw results, which are presented as an HTML page to extract the actual predicted gene and peptide sequences. This is done through the `Helper` class within the `org.jfb.util` package. We have created a separate package for this to enable developers to use this code in a different application that requires similar functionality without the need to extract it from the main application or block of code (**Listing 5.8**).

Listing 5.8. `org.jfb.util package`

```
package org.jfb.util;

import org.jfb.swinggenscan.GenScanResult;
```

```
import java.util.ArrayList;
import java.util.Collection;

public class Helper {
    public                    static                    GenScanResult
extractPeptideAndGene(String rawHtml) {
        final    String    begin    =    "Predicted    peptide
sequence(s):";
        final String end = "<b>Explanation</b>";
        String                allSequences                    =
rawHtml.substring(rawHtml.indexOf(begin)   +   begin.length(),
rawHtml.indexOf(end));
        if (allSequences.indexOf("NO PEPTIDES PREDICTED") >
0) {
            return new GenScanResult();
        }
        int beginIndex = allSequences.indexOf('>');
        allSequences  =  allSequences.substring(beginIndex  +
1, allSequences.length());
        beginIndex = allSequences.indexOf('>');
        allSequences   =   allSequences.substring(beginIndex,
allSequences.length());
        final String[] results = allSequences.split("\n");
        Collection sequences = new ArrayList();
        StringBuffer sb = new StringBuffer();
        for (int i = 0; i < results.length; i++) {
            final String line = results[i];
            if (line.trim().length() == 0) {
                sequences.add(sb.toString());
                sb = new StringBuffer();
            } else {
                sb.append(line).append("\n");
            }
        }
        sequences.add(sb.toString());
        String[] res = new String[sequences.size()];
        sequences.toArray(res);
        final GenScanResult result = new GenScanResult();
        result.setPeptideAndGene(res);
        return result;
    }
}
```

Running SwingGenScan

Fig. 5.11 to **Fig. 5.14** demonstrate a typical run of the `SwingGenScan` application beginning with the pasting of a sequence – in this case – the complete sequence of the human chromosome number 8 (GI number 24850538) and the printing of the predicted genes and peptides.

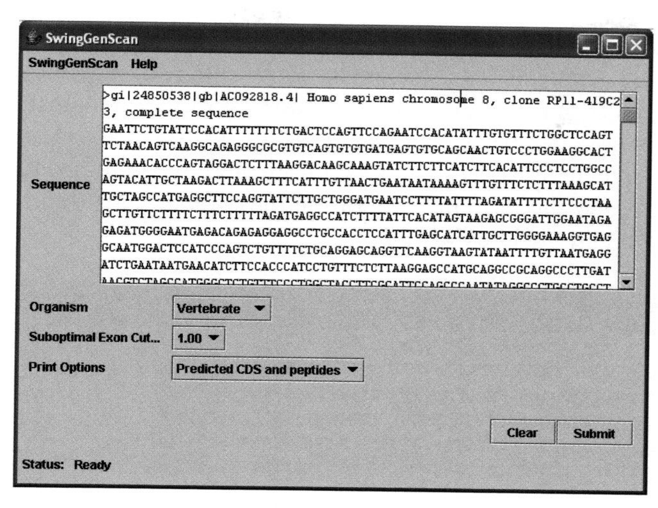

Fig. 5.11. Running SwingGenScan

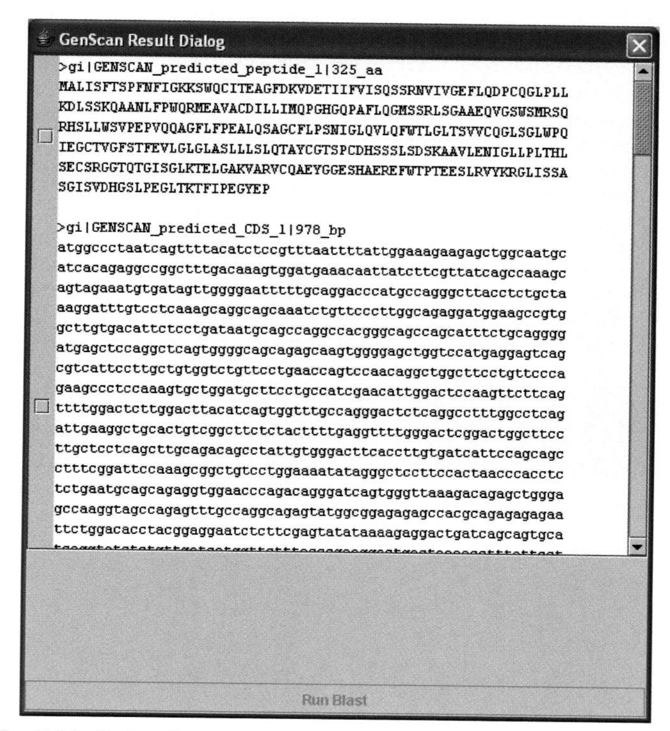

Fig. 5.12. SwingGenScan results

The "`Run Blast`" button remains disabled as long as no sequence is selected for BLAST analysis and becomes active after a sequence is selected (**Fig. 5.12** and **Fig. 5.13**). **Fig. 5.13** and **Fig. 5.14** further demonstrate how predicted sequences can be selected and sent for further analysis using BLAST. Note that selected sequences can be unselected by simultaneously pressing the Control and the left click button on the Mouse (on Windows) and the Apple button and the click (on Mac).

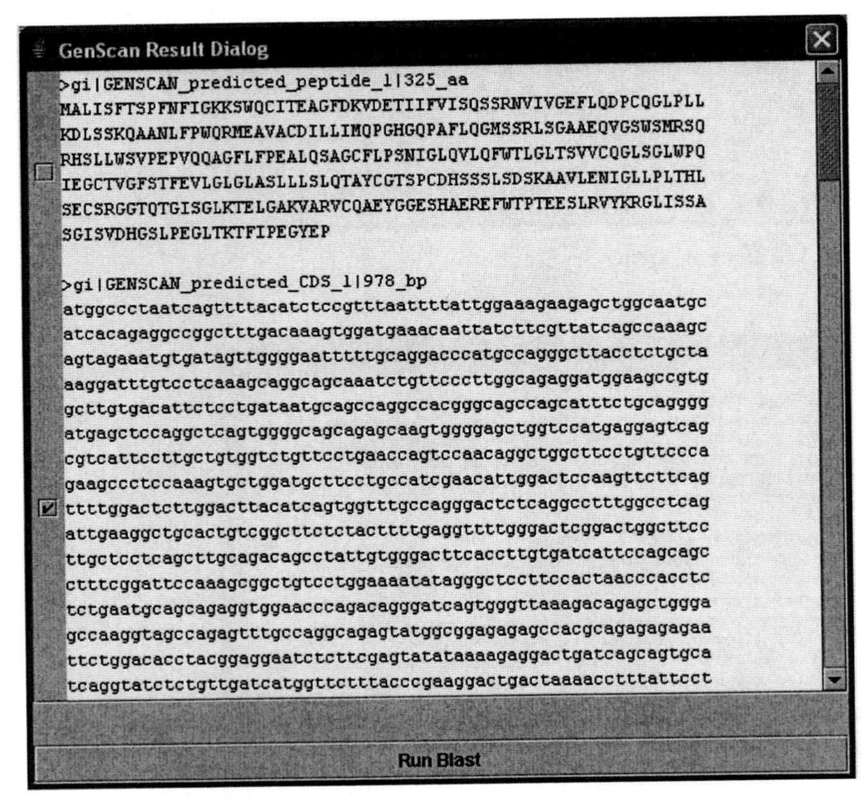

Fig. 5.13. Selecting sequences for BLAST analysis

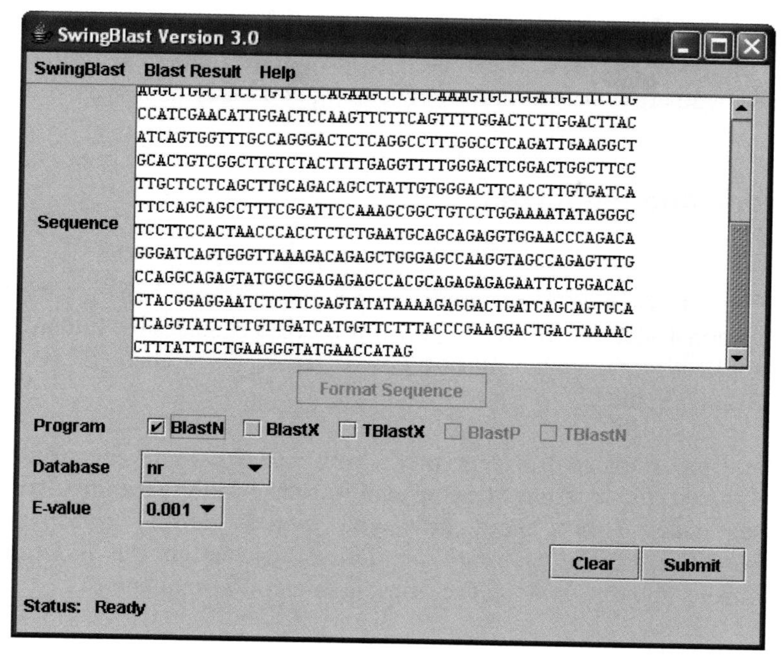

Fig. 5.14. Sending predicted genes to SwingBlast for BLAST analysis

Only BLASTN has been implemented in the `SwingGenScan` application for the purpose of demonstration. The user can further develop the application by adding functionality for other BLAST operations. The Genscan-BLAST analysis pipeline can be implemented in a completely different manner than described here. For example, the `Genscan` output window displaying the gene and peptide predictions can be modified to contain the appropriate widgets to perform multiple BLAST analyses on multiple selected sequences without the intermediate step of invoking the `SwingBlast` application. The implementation shown here is one of many ways to achieve the same end-result.

Summary

In this Chapter, we have demonstrated how we can create a basic gene prediction and annotation pipeline by connecting the Genscan and BLAST programs together. We created the BLAST application separately and tied it together with Genscan thereby building an analytic pipeline that demonstrates reuse of existing code libraries. The addition of functionality to Genscan to enable BLAST analysis of predicted sequences is an example of a real-life use case that will have much practical utility for researchers who are involved in the sequencing and study of new genomes.

Questions and Exercises

1. The `SwingGenScan` application created in the Chapter demonstrated the ability to perform BLASTN searches. Extend the application to enable other types of BLAST searches (BLASTX, BLASTP, etc.).

2. An important goal of gene prediction is to decipher gene structure – that is, the location of exons and introns – in the input nucleotide sequence. Think about how you would identify intron-exon boundaries from Genscan predictions and align the individual introns and exons along the original nucleotide sequence.

Additional Resources

- GenomeScan - http://genes.mit.edu/genomescan.html
- Glimmer - http://www.cbcb.umd.edu/software/glimmer/
- HMMGene - http://www.cbs.dtu.dk/services/HMMgene/
- TwinScan - http://genes.cs.wustl.edu/

Selected Reading

Prediction of complete gene structures in human genomic DNA. Burge, C. and Karlin, S. (1997) J. Mol. Biol. 268, 78-94.

Finding the genes in genomic DNA. Burge, C. B. and Karlin, S. (1998) Curr. Opin. Struct. Biol. 8, 346-354.

Computational inference of homologous gene structures in the human genome. Yeh, R.-F., Lim, L. P., and Burge, C. B. (2001) Genome Res. 11: 803-816.

Improved microbial gene identification with GLIMMER (1999) A.L. Delcher, D. Harmon, S. Kasif, O. White, and S.L. Salzberg. Nucleic Acids Research 27:23, 4636-4641.

Two methods for improving performance of an HMM and their application for gene finding. In Proc. of Fifth Int. Conf. on Intelligent Systems for Molecular Biology, ed. Gaasterland, T. et al., Menlo Park, CA: AAAI Press, 1997, pp. 179-186.

Chapter VI

cancer Biomedical Informatics Grid (caBIG™)

cancer Biomedical Informatics Grid

Whole genome sequencing projects that led to the sequencing and assembly of the human genome and scores of other vertebrate and invertebrate genomes have changed the face of biology and medicine forever. The convergence of molecular-scale biological science, high-throughput technologies and large-scale computing has led to an explosive growth in the volume of information that is available to the modern day biomedical scientist. The success of biomedical research in designing effective therapies for the treatment of complex diseases such as cancer is fundamentally dependent on our ability to integrate and assimilate this raw and largely unstructured data from a variety of experimental platforms encompassing the genomics, proteomics, transcriptomics and the pharmacological and clinical domains. It is also increasingly becoming evident that cooperation among research organizations across geographical boundaries and an open sharing of datasets and analytic tools as well as individual expertise and knowledge is critical to the continued advancement of biomedical research towards its goals.

The cancer Biomedical Informatics Grid project or caBIG™ (pronounced see-ay-big) is built on this very premise. We had provided an introduction to the caBIG™ program in Chapter 1. To recap, the caBIG project was launched in July 2003 and is initiated and funded by the United States National Cancer Institute (NCI) under the aegis of the United States National Institutes of Health (NIH). CaBIG™ is a critical component of NCI's challenge goal of eliminating suffering and death due to cancer by the year 2015. Indeed, CaBIG™ is an effort designed to achieve a level of cross-disciplinary integration that is unprecedented in the history of cancer research. According to NCI Director, Dr. Andrew von Eschenbach, "*...caBIG will become the 'World Wide Web' of cancer research informatics and will accelerate the development of exciting discoveries in all areas of cancer research*". According to the official website (http://cabig.nci.nih.gov/), caBIG is a voluntary, open source, open access initiative that is being designed and built in partnership with the cancer research community across the United States. Since the caBIG pilot program was launched, more than 50 interested NCI-designated cancer centers and more than 800 individuals have participated in the development of the vision, approach and structure of caBIG..

Structure and Organization of caBIG™

caBIG™ participating institutions are organized into Workspaces that are devoted to specific domains of interest relevant to cancer research. Currently, there are four Domain Workspaces, two Cross Cutting Workspaces and three Strategic Level Workspaces. **Table 6.1** provides names and descriptions of the various Workspaces under caBIG™.

Table 6.1. Structure of caBIG™

Workspace name	Purpose
Domain Workspaces	
Clinical Trial Management Systems Workspace	Modular development of tools for the management of clinical trials. These include development of a structured model for protocol representation as well as tools for managing and reporting adverse events that occur during the course of a clinical trial, a laboratory interface module to facilitate automated submission of data to clinical trials systems, a reporting module to submit data electronically to NCI's CDUS (Clinical Data Update System) and the NCI's Clinical Trial Monitoring Service (CTMS) and a financial/billing module to monitor budgets and expenditure in clinical trials. The Workspace is divided into special interest groups for each of these difference activities.
Integrative Cancer Research	Development of modular and interoperable tools and

Workspace	interfaces that provide for integration of clinical and basic research data derived from genomics and proteomics platforms. The Workspace is organized into special interest groups devoted to topics such as Genome Annotation, Microarray Repositories, Pathways Tools, Data Analysis & Statistics, Population Sciences and Translational Tools. Tools being developed under the Workspace include Rproteomics (MALDI-TOF proteomics analysis tool), Gene Ontology Miner (tool for aggregate analysis of gene sets), HapMap (map of haplotypes in human genome), caArray (cancer microarray data management system), Distance Weighted Discrimination (microarray data analysis integrator), Visual and Statistical Data Analyzer (multivariate statistical visualization tool for the analysis of complex data), FunctionExpress (integrated analysis and visualization of microarray data), Quantitative Pathway Analysis in Cancer (pathway modeling and analysis tool), TrAPSS (disease gene mutation discovery and analysis tool), etc.
In Vivo Imaging Workspace	Development of tools to share and integrate the wealth of information provided by in vivo imaging with other types of data. The in vivo imaging technologies and modalities will include systems for research and clinical imaging of live patients and animals (including single-cell organisms) used as model systems for human disease.
Tissue Banks and Pathology Tools Workspace	Development and integration of tissue bank and pathology tools and infrastructure components to enable researchers to locate and analyze tissue specimens for use in cancer research based on tissue, clinical, and genomic characteristics. Tools created under this Workspace include a standard Biospecimen Object Model and suite of tools to facilitate specimen management, annotation and sharing. Specific applications being developed are a specimen inventory and tracking system (caTISSUE Core), a mapping module to get data from tumor registries and clinical anatomy laboratory information systems (caTISSUE Clinical Annotation Engine) and a cancer Text Information Extraction System to automate the process of coding, storing and retrieving data from free-text Pathology Reports (caTIES).

Cross Cutting Workspaces	
Architecture Workspace	Development of tools to ensure consistent application of caBIG™ principles by the large caBIG™ developer community and to meet the caBIG™ program goals of data sharing and interoperability on the grid. Activities include formulating guidelines and definitions for caBIG™ participants to evaluate the maturity level of potential caBIG™ systems and applications (caBIG™ Compatibility Guidelines), development of the grid infrastructure to support the caBIG™ community (caGrid), development of a comprehensive grid security infrastructure for managing federated authentication and authorization in caBIG™, etc.
Vocabularies and Common Data Elements Workspace	Development of policies and guidelines to evaluate and integrate systems based on vocabulary and ontology content as well as software systems for content delivery. Among the major deliverables of this Workspace are the Common Data Elements (CDE) Governance Model to manage the development and administration of CDEs in the Domain Workspaces, data standards approval guidelines for defining the procedures for reviewing and approving data standards, procedures for review and approval of new VCDE content to provide for overall standardization of CDEs within caBIG™, a vocabularies deployment document which lists vocabulary standards consistent with caBIG™ compatibility requirements and LexGrid, a vocabulary server that can be accessed through a well-structured application programming interface (API) capable of accessing and distributing vocabularies as commodity resources.
Strategic Level Working Groups	
Strategic Planning Working Group	Development of strategic planning and vision guidelines in support of the caBIG™ Oversight Board. Activities include creating white papers and planning documents that help define the strategic goals for each individual Workspace as well as for the overall caBIG™ project, along with metrics to measure the success of defined objectives.
Data Sharing and Intellectual Capital	Development of policies and white papers to clarify caBIG's stand on issues surrounding data sharing and intellectual property. Some of the major activities of

	the Working Group include development of guidelines and a model agreement for use by caBIG™ participant institutions to distribute caBIG™ software and related documentation, a caBIG™ publications policy, guidelines on best practices and model agreements for the sharing of data and of biospecimens, reagents and other materials, and a white paper on the de-identification of patient data.
Training Working Group	Development of a caCORE curriculum designed to prepare caBIG™ participants to operate and use the NCI resources such as Enterprise Vocabulary Services (EVS), Cancer Data Standards Repository (caDSR), and Cancer Bioinformatics Infrastructure Objects (caBIO) as well as creating templates and guidelines for caBIG™ documentation and training and organizing boot camps to impart training on caBIG™ technologies.

Further details on caBIG™, its constituent Workspaces and Working Groups and their objectives are available on the WWW at http://cabig.nci.nih.gov/. The ultimate aim of caBIG™ is to enable researchers to collect comprehensive data about cancer in a standardized manner, to enable the study of cancer data as a whole, thereby accelerating the pace of cancer research.

The purpose of this chapter is to not only inform the readers of current efforts in the area of cancer research but also provide knowledge about the technologies that are being developed as an integral part of the effort so that the biomedical and the computer scientists among us can begin using them and in so doing, contribute to their continued development that will ultimately lead to better healthcare solutions and better care and treatments for patients. We will begin by reviewing a few tools and technologies that are relevant to our understanding of how information technologies can assist biomedical research.

Data Integration and ETL

Biomedical researchers routinely need to access and cross-reference sequence and related annotation data from a wide variety of sources such as *PubMed*, *Entrez Gene* (previously called *LocusLink*), *Gene Ontology*

(GO), *UniGene*, *Swiss-Prot*, *Ensembl*, *HomoloGene*, *UniSTS*, etc. Because difference data sources use different formats, it is not easy to compare and combine data from these sources unless they are converted into a common format. Data in UniGene, for example, is presented in text format; data in the GO database is described in an XML format and data in Entrez Gene is available in binary Abstract Syntax Notation number One (ASN.1) format.

CaBIG™ also handles a wide array of data sources, types and formats, from a number of different public domain sources since one of its major goals is to enable access to and sharing of translational research data between cancer researchers. In order to facilitate integration of diverse data types, tools that perform what is known as *Extract*, *Transform* and *Load* (ETL) functions are used. These tools convert data in different formats into a common, standard, usable format. The first step - Extraction - is the part that establishes access to the external database or source that contains the data of interest. The next step - Transformation – analyzes the original data format and converts it to fit with the format of the target repository. For example, information on a gene id can be coded as an XML tag in the form:

<gene id="<my id>"/>

or, as an SQL varchar(64) which means a string of variable length with a maximum size of 64 characters, and other formats. When we design the ETL strategy, we will first create business rules and define the format of the gene identifier that will be used to store that information in the target repository. If this is the SQL varchar(64) type, we will transform data from sources that use a different format into this pre-selected target gene ID format.

The Transformation step can also involve a data-cleansing step to eliminate bad or duplicate entries from input data sources. This process can be done after transforming the data or just before adding data to the target repository. The last step - Load – gets the transformed data loaded into a data repository or a data warehouse, which is optimized to enable faster access to the stored data. An additional step after Extract-Transform-Load is Transportation, which facilitates transport of the formatted data from its current location to the defined location, before it is processed or used further.

A number of open source ETL tools are available, for example, Kettle (available from http://www.kettle.be/), Octopus (available from http://www.enhydra.org/tech/octopus/index.html), and others. Examples of ETL tools being developed under the caBIG™ program include *cancer Function Express tool* (caFE), which annotates individual *probe* sequences (short DNA sequences that represent individual genes or transcripts from a particular genome) on *microarray chips* (arrays of thousands of individual probe sequences embedded on a substrate to detect the presence of specific genes or transcripts in a given genome by hybridizing probes with nucleic acids from the test sample) using data from a number of NCBI and other public databases.

cancer Common Ontologic Representation Environment (caCORE)

A critical component of the partnership between NCI and the Cancer Centers in building the biomedical informatics grid is NCI's Center for Bioinformatics (NCICB). NCICB's mission is to create a close knit and cooperative cancer research community and an interoperable federation of informatics resources covering all aspects of cancer research. NCICB is providing critical support for caBIG through the development of *caCORE*, an open source semantic enterprise architecture for NCI-supported research information systems for genomic and clinical research. A large number of NCI applications such as the *Cancer Molecular Analysis Project* (CMAP), the *Cancer Models Database* (caMOD), and *Gene Expression Data Portal* (GEDP) are directly supported by caCORE. A list of publicly available data sources in the caCORE database is provided in **Table 6.2**. More information on caCORE is available on the NCI caCORE.

Table 6.2. caCORE data sources

Name	Purpose
CGAP Cancer Genome Anatomy Project (CGAP)	Determine the gene expression profiles of normal, precancer, and cancer cells, leading eventually to improved detection, diagnosis, and treatment for the patient.
CGAP Genetic Annotation Initiative (GAI)	Develop a systematic and comprehensive notation of variations in the DNA sequences of each cancer-related gene.
Mouse Models of Human Cancers Consortium	Derive and characterize mouse models, and to generate resources, information, and innovative

(MMHCC)	approaches to the application of mouse models in cancer research.
Cancer Molecular Analysis Project (CMAP)	Facilitate the identification and evaluation of molecular targets in cancer by integrating comprehensive molecular characterizations of cancer.
Gene Expression Data Portal (GEDP)	Provide access to microarray data as well as online data annotation and analysis tools.
Integrated Molecular Analysis of Genomes and their Expression (IMAGE) Consortium	Establish a common resource of publicly available cDNA libraries for access to sequence, map, and expression data.

caCORE is built on the principle of *Model Driven Architecture*, which is a way to organize and manage enterprise architectures supported by automated tools and services for both defining the models and facilitating transformations between different types of models. CaCORE is built on an *n-tier architecture* model and provides *open source Application Programming Interfaces (APIs)* which allow for easy access to data by applications.

The main components of caCORE are:

- *Enterprise Vocabulary Services* (EVS): Controlled vocabulary resources (such as the *NCI thesaurus* and *metathesaurus*) for the life sciences domain that provide a context driven semantic basis for the construction of data elements, classes, and objects.

- *Cancer Data Standards Repository* (caDSR): A metadata registry based upon the ISO/IEC11179 standard that renders research data on cancer reusable and interoperable. The 11179 standard created by the International Organization for Standardization (ISO) and the International Electrotechnical Commission (IEC) specifies the criteria for metadata that are necessary to describe data, as well as the management and administration of that metadata in a metadata registry.

- Cancer Bioinformatics Infrastructure Objects (caBIO): A suite of software, vocabulary, and metadata models for cancer research.

We will explore caBIO objects in further detail in this chapter.

Cancer Bioinformatics Infrastructure Objects (caBIO)

caBIO objects constitute the primary programming interface to caCORE. caBIO objects are implemented using Java and *Java Bean* technology, and model the behavior of hierarchies of biological entities such as *genes*, *sequences* and *chromosomes*, their constituent molecular forms such as *Single Nucleotide Polymorphisms* (SNPs, a single nucleotide difference at a defined location within an individual's DNA sequence), and other entities such as *clones*, *libraries*, *agents*, *pathways*, *tissues* and diseases. A representative list of objects and their descriptions are shown in **Table 6.3**.

Table 6.3. caBIO domain objects

Object name	Description
Gene	The basic physical and functional unit of heredity. Gene objects are the effective portal to most of the genomic information provided by the caBIO data services such as organs, diseases, chromosomes, pathways, sequence data, and expression experiments.
GeneAlias	An alternative name for a gene; provides descriptive information about the gene (as it is known by this alias), as well as access to the Gene object it refers to.
GoOntology	An object providing entry to a Gene object's position in the Gene Ontology Consortium's controlled vocabularies. GoOntology provides access to gene objects corresponding to the ontological term, as well as to ancestor and descendant terms in the ontology tree.
Target	A gene thought to be at the root of a disease etiology and targeted for therapeutic intervention. Defined and used by the CMAP project.
Protein	An object representation of a protein; Protein objects provide access to the encoding gene via its GenBank ID, the taxon in which this instance of the protein occurs, and references to homologous proteins in other species.
Disease	Specifies a disease name and ID; also provides access to ontological relations to other diseases; clinical trial protocolstreating the disease; and specific histologies associated with instances of the disease.
Pathway	An object representation of a molecular/cellular pathway compiled by BioCarta. Pathways are associated with specific Taxon objects, and contain multiple Gene objects, which may be targets for treatment.

Therapeutic agent	A therapeutic agent (drug, intervention therapy) used in a clinical trial protocol.
ClinicalTrialProtocol	The protocol associated with a clinical trial; organizes administrative information about the trial such as Organization ID, participants, phase, etc. provides access to the administered Agents.
Histopathology	An object representing anatomical changes in a diseased tissue sample associated with an expression experiment; captures the relationship between organ and disease.

caBIO provides programmatic access to a variety of open source genomic, biological, and clinical data sources available from the NIH such listed previously (such as Unigene, EntrezGene, etc.) as well as others such as *Biocarta* and clinical trials protocols, etc. caBIO is built upon open source technologies such as Java, *Simple Object Access Protocol* (SOAP, an XML based platform and language independent protocol for exchanging information between applications over the web), *Apache*, *Jakarta Tomcat*, *XML* and *UML*. There are a number of ways that users can access caBIO. Java-based clients communicate with caBIO via the Java API, which contains the domain objects provided by the caBIO.jar file. Non-Java based applications can communicate via SOAP, or by using the caBIO HTTP API and receive objects as XML. caBIO provides access to curated data from multiple sources as described in **Table 6.4**.

Table 6.4. caBIO data sources

NCBI UniGene	Unigene provides a nonredundant partitioning of the genetic sequences contained in GenBank into gene clusters. Each such cluster has a unique UniGene ID and a list of the mRNA and EST sequences that are subsumed by that cluster.
NCBI Entrez Gene (previously called LocusLink)	Entrez Gene contains curated sequence and descriptive information associated with a gene such as gene name, aliases, sequence accession numbers, phenotypes, UniGene cluster IDs, OMIM IDs, gene homologies, associated diseases, map locations, etc.
Gene Ontology (GO) terms	The Gene Ontology Consortium provides a controlled vocabulary for the description of molecular functions, biological processes, and cellular components of gene products.
NCBI HomoloGene	HomoloGene is a resource for curated and calculated gene homologs.
BioCarta pathways	BioCarta provides detailed graphical renderings of

	pathway information concerning apoptosis, cell signalling, cell cycle regulation, immunology, metabolism, and neuroscience, etc.
NCI Cancer Therapy Evaluation Program (CTEP)	CTEP funds an extensive national program of basic and clinical research to evaluate new anti-cancer agents, with a particular emphasis on translational research to elucidate molecular targets and drug mechanisms.
NCI Cancer Models Database (caMOD)	caMOD provides information on animal models of human cancer.

Downloading and Configuring caBIO

caBIO can be downloaded from the NCICB website at:

```
http://ncicb.nci.nih.gov/download/index.jsp
```

Download caCORE3-1.zip file (or the latest available version), unzip to extract the required libraries and save them in an appropriate location making sure that the absolute path to client.jar is declared in the Java classpath. **Fig. 6.1** below shows the caCORE download page.

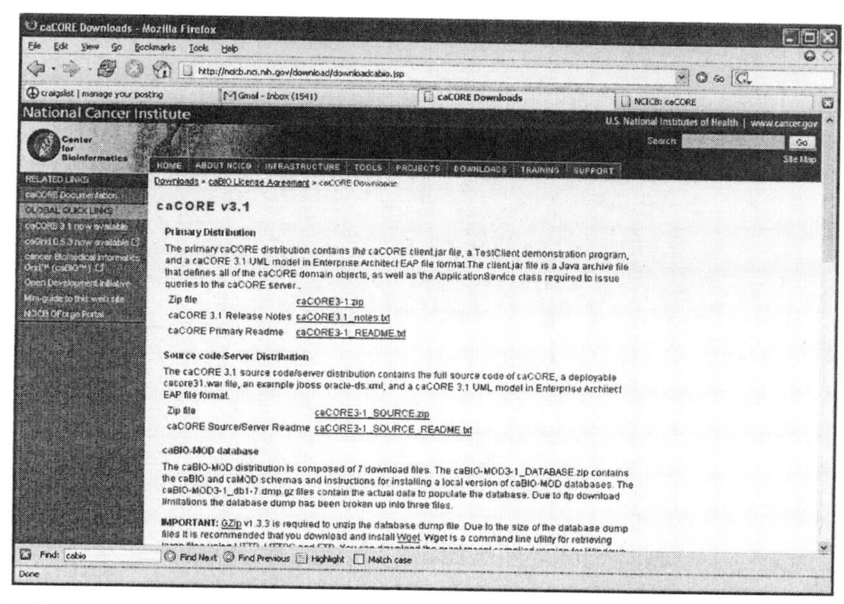

Fig. 6.1. caBIO download page

Now that we have reviewed some of the concepts, technologies and resources available to us from NIH, NCI and other sources, we will create a simple practical application to demonstrate how to integrate the individual isolated bits of data together into a richer, more usable dataset.

Creating the JcaBIO Application

We will create an application based on the caBIO API that we will call JcaBIO to demonstrates how data pertaining to the *Gene* and the *Agent* object can be retrieved using caBIO API. We will create three search functions as outlined below that will define the business logic of the application:

Gene search function: The gene search function will create a report that provides information such as gene name and symbol, Unigene Cluster ID, associated GO terms, gene product name and aliases.

Pathway search function: The pathway search function creates a report that provides information on the pathways that a gene participates in along with a description and a link to the pathway map on BioCarta.

Agent search function: The agent search function creates a report that contains the names of the target(s) that a therapeutic agent binds, the clinical trials that an agent is involved in along with the status, Phase and the name of the institution conducting the trial.

According to this scheme, we will need four command buttons – one each for creating the Gene, Pathway and Agent report and one to clear the report. We will label the command buttons, "Run a Gene Search", "Run a Gene/Pathway Search" and "Run an Agent Search" respectively. We will need a text area to display the reports. We will place this below the command buttons. We will need one text box each to enter the gene name, the agent name and specify the number of reports we want to retrieve for each search. We will place a default value of 10 in the last text box to begin with. We also need a message area to provide the users information on the current state of the application. When the application is launched and when a search is complete, the status will display the "`Ready!`" message. We will place the status bar below the text area.

When the application is initially launched, all the command buttons will be disabled; the command buttons will become available after a valid gene or an agent is entered into the appropriate fields. Only the appropriate command buttons corresponding to the entries will be activated. An entry in the Gene field, for example, will activate the "Run a Gene Search" and "Run a Gene/Pathway Search" buttons while an entry in the Agent field will activate the "Run an Agent Search" button. The Clear button will be activated only after a search has been run and there are results to display.

JcaBIO Classes and Application Structure

The structure of the `JcaBIO` application is shown in **Fig. 6.2**.

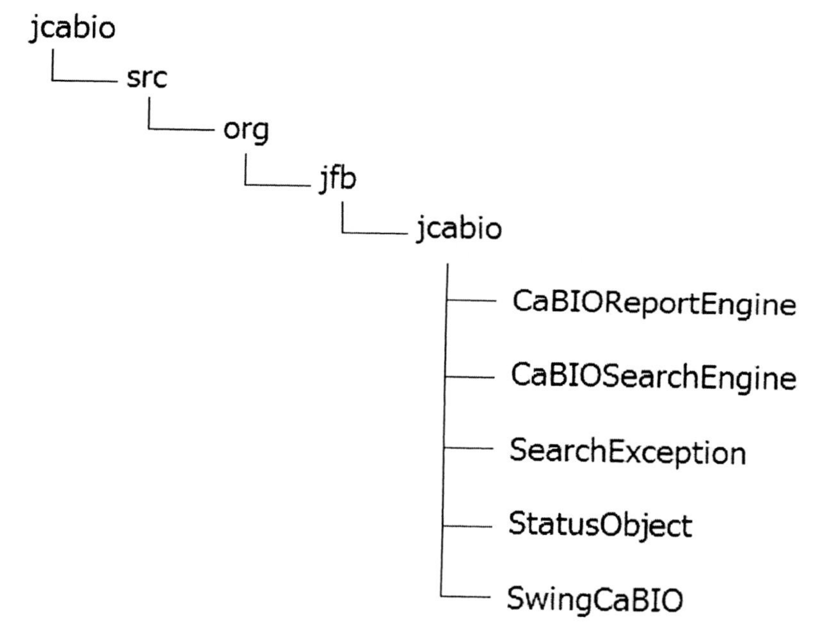

Fig. 6.2. Structure of JcaBIO

A description of the various classes and the corresponding code is as follows:

`SwingCaBIO`: This is the main *Swing* application interface that enables users to send queries and display reports about genes or agents using the caCORE API.

`SearchException`: This class handles exceptions when a search fails.

`StatusObject`: This class stores information on the state of the `CaBIOSearchEngine` or the `CaBIOReportEngine`, which respectively handle the search and the report processes. Using a `StatusObject` instead of a *String* object affords a more generic way of passing the required information. As a result, we have the freedom to modify the `StatusObject` class without having to change the signature of the method that uses this object. We would need to modify only the content of the code as appropriate.

`CaBIOReportEngine`: This is the class that generates the Gene or Agent reports.

`CaBIOSearchEngine`: This class provides the functionality that enable users to perform gene or agent searches.

The application at start up showing the various Swing components is shown in **Fig. 6.3**.

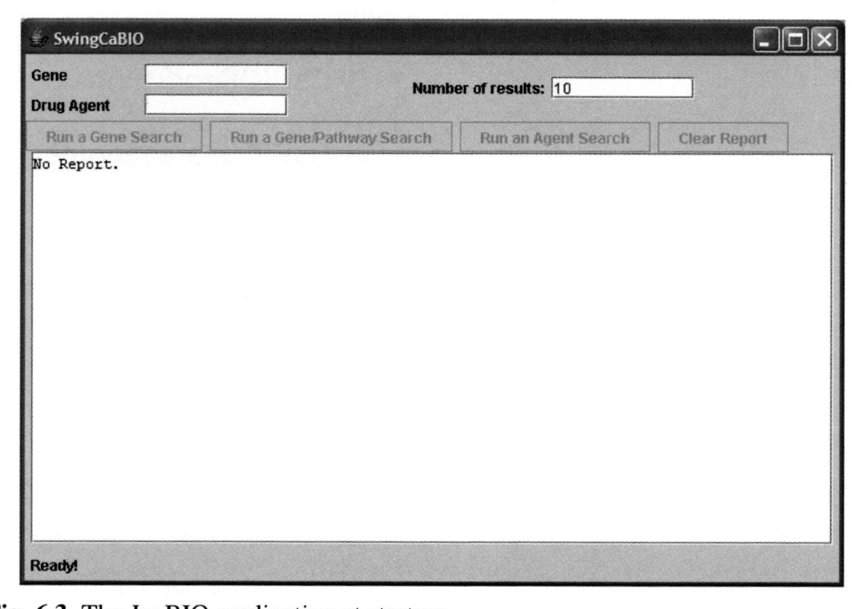

Fig. 6.3. The JcaBIO application at start up

Coding the SwingCaBIO Application

`SwingCaBIO` defines the application interface that the user interacts with to send and retrieve queries using the caCORE API. `SwingCaBIO` is based on the same *Swing* elements and concepts that were described in Chapter 1. `SwingCaBIO` extends *JFrame* in order to generate a basic container for the application. `SwingCaBIO` also extends *DocumentListener* to listen to the *JTextField* objects in order to enable or disable the corresponding buttons that run the report. The constructor `SwingCaBIO()` calls the *super* constructor and adds the *observer* to the Agent search and the report engine. At that point, the application consists of a frame and nothing else built inside. We then call the `init()` function explained below that will build our form for the Gene and Agent searches.

As described earlier, we need three command (search) buttons for running the three custom reports and a text area to display the search results. We use what are called *factory* methods to create the different pieces that are assembled in the `init()` method. *Factory* methods refer to the *Factory Design Pattern*, which specifies a way to create objects without having to know how they are created or assembled. This design allows the developer to change the way the buttons and other Swing components are displayed without interfering with the rest of the Swing components that make the application. The `SwingCaBIO()` method is described below:

```
public SwingCaBIO() throws HeadlessException {
  super();
  AGENT_SEARCH.addObserver(observer);
  REPORT_ENGINE.addObserver(observer);
}

private void init() {
  setTitle("SwingCaBIO");
  final Container contentPane = getContentPane();
  contentPane.setLayout(new BorderLayout());
  JPanel formPanel = createForm();
  JPanel reportPanel = createReportPane();
  statusBar = new JLabel(STATUS_READY);
  statusBar.setBorder(BorderFactory.createEmptyBorder(5, 5,
  5, 5));
  contentPane.add(formPanel, BorderLayout.NORTH);
  contentPane.add(reportPanel, BorderLayout.CENTER);
  contentPane.add(statusBar, BorderLayout.SOUTH);
  setDefaultCloseOperation(JFrame.EXIT_ON_CLOSE);
  pack();
  setSize(DIMENSION);
```

```
    final Dimension screenSize = Toolkit.getDefaultToolkit().
    getScreenSize();
    setLocation(new Point((screenSize.width - SW_WIDTH) / 2,
    (screenSize.height - SW_HEIGHT) / 2));
    show();
  }
```

The `getNumberOfobjectsForResult()` method handles the total number of results to return for a search. As indicated earlier, we will place a default value of 10 for this.

```
  private static final StatusObject STATUS_REPORT_GENERATED =
    new StatusObject("Report generated!", 10);
    final int len =
    getNumberOfObjectsForResult(genes.length);
    sb.append("Search Results for  '" + genePattern + "' (" +
len + " gene(s) found):\n\n");
    for (int i = 0; i < len; i++) {
      Gene gene = genes[i];
      REPORT_ENGINE.printFullGeneReport(gene, sb, i + 1);
    }
```

`We use the showReport()` method to set the text area with the current *StringBuffer* object containing the report generated. Anytime a Swing object is modified, we need to make sure that the method runs in the *event-dispatching thread* (also called the *AWT thread*) to avoid painting problems. The method checks if we are already in the *event-dispatching thread* before calling the `runnable` object.

```
    private void showReport(final StringBuffer sb) {
        Runnable runnable = new Runnable() {
            public void run() {
                jTextArea.setText(sb.toString());
            }
        };
        if (SwingUtilities.isEventDispatchThread())
            runnable.run();
        else
            SwingUtilities.invokeLater(runnable);
    }
```

The methods `insertUpdate()`, `changeUpdate()` and `removeUpdate()` are methods from the `DocumentListener` interface. We will use these methods to update the buttons according to the values found in the gene and the agent fields.

```
  public void insertUpdate(DocumentEvent event) {
```

```
        updateButtons();
    }

    private void updateButtons() {
        boolean enabled = gene.getText().trim().length() > 0;
        runGenePathwayReport.setEnabled(enabled);
        runFullGeneReport.setEnabled(enabled);
        runTargetAgentReport.setEnabled(agent.getText().trim().
        length() > 0);
    }
```

The three update methods are delegating the treatment of the event to the `updateButtons()` method. The `updateButtons()` method enables or disables buttons according to the status of the search and the corresponding report that is generated. We add a utility method called `errorDump()` to create an error message and update the status bar to alert user about any problems encountered:

```
    private void errorDump(StringBuffer sb, SearchException e)
  {
        sb.delete(0, sb.length());
        sb.append("An error occured!\n\n" +
          e.getEmbedded().getMessage());
        updateStatus(new StatusObject("An error occured!", 5));
    }
```

We include a method to update the status of the search and reporting using the `updateStatus()` method. `updateStatus()` sets the text in the status bar with information on the state of the search. We need to invoke and display the update right away to ensure that the user is alerted as soon as an issue arises. For this reason we have implemented the `invokeAndWait()` instead of the `invokeLater()` method.

These methods force the JVM to invoke the `run()` method of the *Runnable* object passed as an argument, inside the *event-dispatching thread.*

```
      private void updateStatus(final StatusObject
statusObject) {
            Runnable runnable = new Runnable() {
                public void run() {

statusBar.setText(statusObject.getStatusText());
                    if (statusObject.hasTimer()) {
                        new Thread(new Runnable() {
                            public void run() {
                                try {
                                    synchronized (this) {
```

```
this.wait(statusObject.getTimer() * 1000);
                                    }
                                    updateStatusToReady();
                                } catch (InterruptedException
e) {
                                    e.printStackTrace();
                                }
                            }
                        }).start();
                    }
                }
            };
            if (SwingUtilities.isEventDispatchThread()) {
                runnable.run();
            } else {
                try {
                    SwingUtilities.invokeAndWait(runnable);
                } catch (InterruptedException e) {
                    e.printStackTrace();
                } catch (InvocationTargetException e) {
                    e.printStackTrace();
                }
            }
        }
```

The `main()` method starts the creation of the `SwingCaBIO` in the *event-dispatching thread*. This avoids having paint methods that freeze the application i.e., the application does not respond to any mouse clicks or keyboard interactions or the application is just gray with no components created in the main frame.

```
public static void main(String[] args) {
  SwingUtilities.invokeLater(new Runnable() {
    public void run() {
      final SwingCaBIO swingCaBIO = new SwingCaBIO();
      swingCaBIO.init();
    }
  });
}
```

The complete code for the `SwingCaBIO` class is provided in **Listing 6.1.**

Listing 6.1. Class SwingCaBIO

```
package org.jfb.jcabio;

import gov.nih.nci.cabio.domain.Agent;
import gov.nih.nci.cabio.domain.Gene;
```

```
import javax.swing.*;
import javax.swing.event.DocumentEvent;
import javax.swing.event.DocumentListener;
import java.awt.*;
import java.awt.event.ActionEvent;
import java.awt.event.ActionListener;
import java.lang.reflect.InvocationTargetException;
import java.util.Observable;
import java.util.Observer;

public class SwingCaBIO extends JFrame implements
DocumentListener {
    private static final int SW_WIDTH = 700;
    private static final int SW_HEIGHT = 600;
    private static final Dimension DIMENSION = new
Dimension(SW_WIDTH, SW_HEIGHT);
    private static final Dimension DIM_FIELD = new
Dimension(85, 18);
    private final static String CABIO_HTTP_SERVER_URL =

     "http://cabio.nci.nih.gov/cacore30/server/HTTPServer";
    private final static ApplicationService APP_SERVICE =
     ApplicationService.getRemoteInstance(CABIO_HTTP_SERVER
_URL);
    private static final CaBIOSearchEngine AGENT_SEARCH =
new CaBIOSearchEngine(APP_SERVICE);
    private static final CaBIOReportEngine REPORT_ENGINE =
new CaBIOReportEngine(APP_SERVICE);
    private static final String STATUS_READY = "Ready!";
    private static final StatusObject
STATUS_REPORT_GENERATED = new StatusObject("Report
generated!", 10);

    private JTextArea jTextArea;
    private JTextField gene;
    private JButton runFullGeneReport;
    private JButton runTargetAgentReport;
    private JButton runGenePathwayReport;
    private JButton clear;

    private JLabel statusBar;
    private JTextField result;
    private JTextField agent;

    public SwingCaBIO() throws HeadlessException {
        super();
        AGENT_SEARCH.addObserver(observer);
        REPORT_ENGINE.addObserver(observer);
    }

    private void init() {
        setTitle("SwingCaBIO");
```

```
            final Container contentPane = getContentPane();
            contentPane.setLayout(new BorderLayout());
            JPanel formPanel = createForm();
            JPanel reportPanel = createReportPane();
            statusBar = new JLabel(STATUS_READY);

statusBar.setBorder(BorderFactory.createEmptyBorder(5, 5, 5,
5));
            contentPane.add(formPanel, BorderLayout.NORTH);
            contentPane.add(reportPanel, BorderLayout.CENTER);
            contentPane.add(statusBar, BorderLayout.SOUTH);
            setDefaultCloseOperation(JFrame.EXIT_ON_CLOSE);
            pack();
            setSize(DIMENSION);
            final         Dimension         screenSize         =
Toolkit.getDefaultToolkit().getScreenSize();
            setLocation(new Point((screenSize.width - SW_WIDTH)
/ 2, (screenSize.height - SW_HEIGHT) / 2));
            show();
        }

        private JPanel createReportPane() {
            jTextArea = new JTextArea();
            jTextArea.setText("No Report.");
            jTextArea.setEditable(false);
            jTextArea.setWrapStyleWord(true);
            jTextArea.setLineWrap(true);
            final Font sf = jTextArea.getFont();
            Font  f  =  new  Font("Monospaced",  sf.getStyle(),
sf.getSize());
            jTextArea.setFont(f);
            jTextArea.getDocument().addDocumentListener(new
DocumentListener() {
                public void insertUpdate(DocumentEvent event) {

clear.setEnabled(jTextArea.getText().trim().length() != 0);
                }

                public void removeUpdate(DocumentEvent event) {

clear.setEnabled(jTextArea.getText().trim().length() != 0);
                }

                public  void  changedUpdate(DocumentEvent  event)
{

clear.setEnabled(jTextArea.getText().trim().length() != 0);
                }
            });
            JPanel jPanel = new JPanel();
            jPanel.setLayout(new BorderLayout());
            final     JScrollPane     jScrollPane      =      new
JScrollPane(jTextArea);
```

```
            jPanel.add(jScrollPane, BorderLayout.CENTER);
            jPanel.setBorder(BorderFactory.createEmptyBorder(0,
 5, 5, 5));
            return jPanel;
        }

        private JPanel createForm() {
            final JPanel genePanel = new JPanel();
            genePanel.setLayout(new        BoxLayout(genePanel,
 BoxLayout.LINE_AXIS));
            gene = new JTextField(10);
            this.gene.setMaximumSize(DIM_FIELD);
            gene.getDocument().addDocumentListener(this);
            final JLabel geneLabel = new JLabel("Gene");
            geneLabel.setPreferredSize(DIM_FIELD);
            genePanel.add(Box.createRigidArea(new  Dimension(5,
 0)));
            genePanel.add(geneLabel);
            genePanel.add(Box.createRigidArea(new  Dimension(5,
 0)));
            genePanel.add(gene);
            genePanel.add(Box.createHorizontalGlue());
            final JPanel agentPanel = new JPanel();
            agentPanel.setLayout(new      BoxLayout(agentPanel,
 BoxLayout.LINE_AXIS));
            agent = new JTextField(10);
            agent.setMaximumSize(DIM_FIELD);
            agent.getDocument().addDocumentListener(this);
            final JLabel agentLabel = new JLabel("Drug Agent");
            agentLabel.setPreferredSize(DIM_FIELD);
            agentPanel.add(Box.createRigidArea(new Dimension(5,
 0)));
            agentPanel.add(agentLabel);
            agentPanel.add(Box.createRigidArea(new Dimension(5,
 0)));
            agentPanel.add(this.agent);
            agentPanel.add(Box.createHorizontalGlue());

            JPanel jPanel = new JPanel();
            jPanel.setLayout(new              BoxLayout(jPanel,
 BoxLayout.PAGE_AXIS));
            jPanel.add(Box.createRigidArea(new     Dimension(0,
 5)));
            jPanel.add(genePanel);
            jPanel.add(Box.createRigidArea(new     Dimension(0,
 5)));
            jPanel.add(agentPanel);
            jPanel.add(Box.createRigidArea(new     Dimension(0,
 5)));

            final JPanel resultPanel = new JPanel();
            resultPanel.setLayout(new    BoxLayout(resultPanel,
 BoxLayout.LINE_AXIS));
```

```
            result = new JTextField("10", 10);
            result.setMaximumSize(DIM_FIELD);
            final JLabel resultLabel = new JLabel("Number of
results:");
            resultPanel.add(Box.createRigidArea(new
Dimension(5, 0)));
            resultPanel.add(resultLabel);
            resultPanel.add(Box.createRigidArea(new
Dimension(5, 0)));
            resultPanel.add(result);

            JPanel jPanelResult = new JPanel();
            jPanelResult.setLayout(new BoxLayout(jPanelResult,
BoxLayout.LINE_AXIS));
            jPanelResult.add(jPanel);
            resultPanel.add(Box.createRigidArea(new
Dimension(20, 0)));
            jPanelResult.add(resultPanel);
            resultPanel.add(Box.createHorizontalGlue());


            runFullGeneReport = new JButton("Run a Gene
Search");
            runFullGeneReport.setToolTipText("Please provide a
gene to search for.");
            runFullGeneReport.setEnabled(false);
            runFullGeneReport.addActionListener(new
ActionListener() {
                public void actionPerformed(ActionEvent event)
{
                    final StringBuffer sb = new StringBuffer();
                    Runnable runnable = new Runnable() {
                        public void run() {
                            final String genePattern =
gene.getText();
                            showReport(new
StringBuffer("Searching with gene '" + genePattern +
"'..."));
                            try {
                                final Gene[] genes =
AGENT_SEARCH.searchGenesWithGenePattern(genePattern);
                                final int len =
getNumberOfObjectsForResult(genes.length);
                                sb.append("Search Results for
'" + genePattern + "' (" + len + " gene(s) found):\n\n");
                                for (int i = 0; i < len; i++) {
                                    Gene gene = genes[i];

REPORT_ENGINE.printFullGeneReport(gene, sb, i + 1);
                                    showReport(new
StringBuffer(jTextArea.getText()).append("\n")
                                            .append("Generated
report for gene '" + gene.getFullName() + "'"));
```

```
                                        if (i + 1 < len)
                                            sb.append("\n\n");
                                    }

updateStatus(STATUS_REPORT_GENERATED);
                                } catch (SearchException se) {
                                    errorDump(sb, se);
                                }
                                showReport(sb);
                            }
                        };
                        new Thread(runnable).start();
                    }
                });
                runGenePathwayReport    =    new    JButton("Run    a
Gene/Pathway Search");
                runGenePathwayReport.setToolTipText("Please provide
a Gene to search for.");
                runGenePathwayReport.setEnabled(false);
                runGenePathwayReport.addActionListener(new
ActionListener() {
                    public void actionPerformed(ActionEvent event)
{
                        final StringBuffer sb = new StringBuffer();
                        Runnable runnable = new Runnable() {
                            public void run() {
                                final    String    genePattern    =
gene.getText();
                                showReport(new
StringBuffer("Searching   with   gene   '"   +   genePattern   +
"'..."));
                                try {
                                    final    Gene[]    genes    =
AGENT_SEARCH.searchGenesWithGenePattern(genePattern);
                                    int              len              =
getNumberOfObjectsForResult(genes.length);
                                    sb.append("Search  Results  for
'" + genePattern + "' (" + len + " gene(s) found):\n\n");
                                    Gene gene;
                                    for (int i = 0; i < len; i++) {
                                        gene = genes[i];

REPORT_ENGINE.printGenePathwayReport(gene, sb, i + 1);
                                        showReport(new
StringBuffer(jTextArea.getText()).append("\n")
                                                .append("Generated
report for gene '" + gene.getFullName() + "'"));
                                        if (i + 1 < len)
                                            sb.append("\n\n");
                                    }

updateStatus(STATUS_REPORT_GENERATED);
                                } catch (SearchException se) {
```

```
                            errorDump(sb, se);
                        }
                        showReport(sb);
                    }
                };
                new Thread(runnable).start();
            }
        });
        runTargetAgentReport = new JButton("Run an Agent
Search");
        runTargetAgentReport.setToolTipText("Please provide
an agent to search for.");
        runTargetAgentReport.setEnabled(false);
        runTargetAgentReport.addActionListener(new
ActionListener() {
            public void actionPerformed(ActionEvent event)
{
                final StringBuffer sb = new StringBuffer();
                Runnable runnable = new Runnable() {
                    public void run() {
                        final String agentPattern =
agent.getText();
                        showReport(new
StringBuffer("Searching with agent '" + agentPattern
                                + "'..."));
                        try {
                            final Agent[] agents =
AGENT_SEARCH.searchAgentsWithAgentPattern(agentPattern);
                            final int len =
getNumberOfObjectsForResult(agents.length);
                            sb.append("Search Results for
'" + agentPattern +
                                    "' (" + len + "
agent(s) found):\n\n");
                            Agent agent;
                            for (int i = 0; i < len; i++) {
                                agent = agents[i];

REPORT_ENGINE.printGeneAgentCliTriReport(agent, sb, i + 1);
                                showReport(new
StringBuffer(jTextArea.getText()).append("\n")
                                        .append("Generated
report for agent '" + agent.getName() + "'"));
                                if (i + 1 < len)
                                    sb.append("\n\n");
                            }

updateStatus(STATUS_REPORT_GENERATED);
                        } catch (SearchException se) {
                            errorDump(sb, se);
                        }
                        showReport(sb);
                    }
```

```
                    };
                    new Thread(runnable).start();
                }
            });
            clear = new JButton("Clear Report");
            clear.addActionListener(new ActionListener() {
                public void actionPerformed(ActionEvent event)
{
                    jTextArea.setText("");
                }
            });
          clear.setEnabled(false);
            JPanel buttonPanel = new JPanel();
            buttonPanel.setLayout(new    BoxLayout(buttonPanel,
BoxLayout.LINE_AXIS));
            buttonPanel.add(runFullGeneReport);
            buttonPanel.add(Box.createRigidArea(new
Dimension(5, 0)));
            buttonPanel.add(runGenePathwayReport);
            buttonPanel.add(Box.createRigidArea(new
Dimension(5, 0)));
            buttonPanel.add(runTargetAgentReport);
            buttonPanel.add(Box.createRigidArea(new
Dimension(5, 0)));
            buttonPanel.add(clear);

            JPanel formPanel = new JPanel();
            formPanel.setLayout(new BorderLayout());
            formPanel.add(jPanelResult, BorderLayout.NORTH);
            formPanel.add(buttonPanel, BorderLayout.CENTER);
            return formPanel;
        }

        private int getNumberOfObjectsForResult(int len) {
            String numOfRes = result.getText();
            if (numOfRes != null && numOfRes.trim().length() >
0) {
                return     Math.min(Integer.parseInt(numOfRes),
len);
            }
            return len;
        }

        private void showReport(final StringBuffer sb) {
            Runnable runnable = new Runnable() {
                public void run() {
                    jTextArea.setText(sb.toString());
                }
            };
            if (SwingUtilities.isEventDispatchThread())
                runnable.run();
            else
                SwingUtilities.invokeLater(runnable);
```

```
        }

      public void insertUpdate(DocumentEvent event) {
            updateButtons();
        }

        private void updateButtons() {
            boolean enabled = gene.getText().trim().length() >
0;
            runGenePathwayReport.setEnabled(enabled);
            runFullGeneReport.setEnabled(enabled);

runTargetAgentReport.setEnabled(agent.getText().trim().length
() > 0);
        }

        private void errorDump(StringBuffer sb, SearchException
e) {
            sb.delete(0, sb.length());
            sb.append("An      error      occured!\n\n"      +
e.getEmbedded().getMessage());
            updateStatus(new StatusObject("An error occured!",
5));
        }

        private Observer observer = new Observer() {
            public void update(Observable observable, Object o)
{
                updateStatus((StatusObject) o);
            }
        };

        private void updateStatusToReady() {
            updateStatus(StatusObject.STATUS_READY);
        }

        private    void    updateStatus(final    StatusObject
statusObject) {
            Runnable runnable = new Runnable() {
                public void run() {

statusBar.setText(statusObject.getStatusText());
                    if (statusObject.hasTimer()) {
                        new Thread(new Runnable() {
                            public void run() {
                                try {
                                    synchronized (this) {

this.wait(statusObject.getTimer() * 1000);
                                    }
                                    updateStatusToReady();
                                } catch (InterruptedException
e) {
```

```
                                    e.printStackTrace();
                                }
                            }
                        }).start();
                    }
                }
            };
            if (SwingUtilities.isEventDispatchThread()) {
                runnable.run();
            } else {
                try {
                    SwingUtilities.invokeAndWait(runnable);
                } catch (InterruptedException e) {
                    e.printStackTrace();
                } catch (InvocationTargetException e) {
                    e.printStackTrace();
                }
            }
        }

        public void removeUpdate(DocumentEvent event) {
            updateButtons();
        }

        public void changedUpdate(DocumentEvent event) {
        }

        public static void main(String[] args) {
            SwingUtilities.invokeLater(new Runnable() {
                public void run() {
                    final    SwingCaBIO    swingCaBIO    =    new
SwingCaBIO();
                    swingCaBIO.init();
                }
            });
        }
    }
```

Coding JcaBIO: The CaBIOReportEngine Class

In order to provide information on what the report engine is doing while it is generating the report, the CaBIOReportEngine class extends `java.util.Observable` to send notification to all *observers* about the status of the generation of the report.

```
public class CaBIOReportEngine extends Observable { }
```

The constructor for the class takes a `gov.nih.nci.system.applicationservice.ApplicationService` object that retrieves further information during report generation as needed.

`CaBIOReportEngine` contains a number of print methods in order to generate the reports. These methods print specific information about the gene or agent (that the user supplied) into the `StringBuffer` object. The `printGene()` method takes two parameters to generate the report – the gene object and the `StringBuffer` object which will contain the information to be included in the report:

```
public void printGene(Gene gene, StringBuffer sb) { }
```

Within the `printGene()` method, we implement methods provided by the caBIO API such as `getFullName()`, `getSymbol()` and `getClusterId()` to access the relevant information about the input gene.

The `printPathways()` method takes the same two parameters to generate the pathways report:

```
    public void printPathways(Gene gene, StringBuffer sb) {
}
```

Information on pathways is obtained as a collection of pathway objects using the method `search()` from the application service object `appService`. The `search()` method requires two parameters - the type of the object we want in the collection result and the gene we need the pathways for as shown below:

```
    final              Collection              tmp              =
appService.search(Pathway.class, gene);
```

Similarly, we use print methods to retrieve information on gene aliases (`printGeneAliases()`), clinical trials (`printClinicalTrials()`), Agent (`printAgent()`) etc. The complete code for `CaBIOReportEngine` is provided in **Listing 6.2**.

Listing 6.2. Class CaBIOReportEngine

```
package org.jfb.jcabio;

import gov.nih.nci.cabio.domain.Agent;
import gov.nih.nci.cabio.domain.ClinicalTrialProtocol;
import gov.nih.nci.cabio.domain.Gene;
import gov.nih.nci.cabio.domain.GeneAlias;
```

```
import gov.nih.nci.cabio.domain.GeneOntology;
import gov.nih.nci.cabio.domain.HomologousAssociation;
import gov.nih.nci.cabio.domain.Pathway;
import gov.nih.nci.cabio.domain.Protein;
import gov.nih.nci.cabio.domain.Target;

import java.text.SimpleDateFormat;
import java.util.Collection;
import java.util.Iterator;
import java.util.Observable;

public class CaBIOReportEngine extends Observable {
    private static final SimpleDateFormat DATE_FORMATTER =
new SimpleDateFormat("yyyy.MM.dd G 'at' HH:mm:ss z");
    private static final StatusObject STATUS_REPORT_DONE =
new StatusObject("Report done!");

    private ApplicationService appService;

    public CaBIOReportEngine(ApplicationService appService)
{
        this.appService = appService;
    }

    public void printGene(Gene gene, StringBuffer sb) {
        sb.append("Name: " + gene.getFullName());
        sb.append("\n-Symbol: " + gene.getSymbol());
        sb.append("\n-Unigene    Cluster    Id:    "    +
gene.getClusterId());
    }

    public void printPathways(Gene gene, StringBuffer sb) {
        notifyObservers(new    StatusObject("Printing    the
pathway report for gene '" + gene.getFullName() + "'..."));
        try {
                final        Collection        tmp        =
appService.search(Pathway.class, gene);
            final int size = tmp.size();
            if (size == 0) {
                sb.append("Gene    not    found    in    any
pathways.");
                notifyObservers(STATUS_REPORT_DONE);
                return;
            }
            sb.append(size + " pathway(s) found: \n");
            Pathway[] pathways = new Pathway[size];
            tmp.toArray(pathways);
            for (int i = 0; i < pathways.length; i++) {
                Pathway pathway = pathways[i];
                sb.append("\t-Pathway    name:    "    +
pathway.getName());
                sb.append("\n\t-Description:    "    +
pathway.getDisplayValue());
```

```
                    sb.append("\n\t-Pathway                    Map:
http://www.biocarta.com/pathfiles/" + pathway.getName() +
".asp");
                    if (i + 1 < pathways.length) {
                        sb.append("\n");
                    }
                }
            } finally {
                notifyObservers(new        StatusObject("Pathway
report done for gene '" + gene.getFullName() + "'!"));
            }
        }

        public  void  printGeneAliases(Gene  gene,  StringBuffer
sb) {
            notifyObservers(new StatusObject("Printing the gene
alias report for gene '" + gene.getFullName() + "'..."));
            try {
                final          Collection          tmp          =
appService.search(GeneAlias.class, gene);
                final int size = tmp.size();
                if (size == 0) {
                    sb.append("No gene aliases found.");
                    notifyObservers(STATUS_REPORT_DONE);
                    return;
                }
                sb.append(size + " gene aliases found: ");
                GeneAlias[] geneAliases = new GeneAlias[size];
                tmp.toArray(geneAliases);
                for (int i = 0; i < geneAliases.length; i++) {
                    GeneAlias geneAlias = geneAliases[i];
                    sb.append(geneAlias.getName());
                    if (i + 1 < geneAliases.length) {
                        sb.append(", ");
                    }
                }
            } finally {
                notifyObservers(new   StatusObject("Gene   alias
report done for gene '" + gene.getFullName() + "'!"));
            }
        }

        private void printAgent(Agent agt, StringBuffer sb) {
            sb.append("Drug Agent Name: " + agt.getName());
            final String source = agt.getSource();
            sb.append("\n-Agent Source: " + (source != null ?
source : "Unknown"));
        }

        public void printGenes(Target target, StringBuffer sb)
{
            notifyObservers(new StatusObject("Printing the gene
report for target '" + target.getName() + "'..."));
```

```
            try {
                final          Collection          tmp          =
appService.search(Gene.class, target);
                final int size = tmp.size();
                if (size == 0) {
                    sb.append("No genes found.");
                    notifyObservers(STATUS_REPORT_DONE);
                    return;
                }
                sb.append(size + " gene(s) found: ");

                Gene[] genes = new Gene[size];
                tmp.toArray(genes);
                for (int i = 0; i < genes.length; i++) {
                    Gene agt = genes[i];
                    printGene(agt, sb);
                    if (i + 1 < genes.length) {
                        sb.append("\n");
                    }
                }
            } finally {
                notifyObservers(new  StatusObject("Gene  report
done for gene '" + target.getName() + "'!"));
            }
        }

        public void printClinicalTrials(Agent agt, StringBuffer
sb) {
            notifyObservers(new   StatusObject("Printing   the
clinical  trial  report  for  agent  '"  +  agt.getName()  +
"'..."));
            try {
                final          Collection          tmp          =
appService.search(ClinicalTrialProtocol.class, agt);
                final int size = tmp.size();
                if (size == 0) {
                    sb.append("No  clinical  trials  found  for
agent.");

                    notifyObservers(STATUS_REPORT_DONE);
                    return;
                }
                sb.append(size + " clinical trial(s) found: ");
                ClinicalTrialProtocol[]  clinicalTrials  =  new
ClinicalTrialProtocol[size];
                tmp.toArray(clinicalTrials);
                for (int i = 0; i < clinicalTrials.length; i++)
{
                    ClinicalTrialProtocol    clinicalTrial    =
clinicalTrials[i];
                    sb.append("\n\nTitle:          "          +
clinicalTrial.getTitle());
                    sb.append("\n-Status:          "          +
clinicalTrial.getCurrentStatus());
```

```
                        sb.append("\n-Date:            " +
DATE_FORMATTER.format(clinicalTrial.getCurrentStatusDate()));
                        sb.append("\n-Lead Organization Name: " +
clinicalTrial.getLeadOrganizationName());
                        sb.append("\n-Phase:           " +
clinicalTrial.getPhase());
                        sb.append("\n-Participation   Type:    " +
clinicalTrial.getParticipationType());
                        if (i + 1 < clinicalTrials.length) {
                            sb.append("\n");
                        }
                    }
            } finally {
                notifyObservers(new      StatusObject("Clinical
trial report done for agent '" + agt.getName() + "'!"));
            }
        }

        public   void   printGeneOntology(Gene   gene,   StringBuffer
sb) {
            notifyObservers(new StatusObject("Printing the gene
ontology report for gene '" + gene.getFullName() + "'..."));
            try {
                final          Collection          tmp           =
appService.search(GeneOntology.class, gene);
                final int size = tmp.size();
                if (size == 0) {
                    sb.append("No associated GO terms found.");
                    notifyObservers(STATUS_REPORT_DONE);
                    return;
                }
                sb.append(size + " GO Term(s) found: ");
                GeneOntology[]     geneOntologies      =      new
GeneOntology[size];
                tmp.toArray(geneOntologies);
                for (int i = 0; i < geneOntologies.length; i++)
{
                    GeneOntology          geneOntology           =
geneOntologies[i];
                    sb.append(geneOntology.getName());
                    if (i + 1 < geneOntologies.length) {
                        sb.append(", ");
                    }
                }
            } finally {
                notifyObservers(new StatusObject("Gene ontology
report done for gene '" + gene.getFullName() + "'!"));
            }
        }

        public void printProteins(Gene gene, StringBuffer sb) {
            notifyObservers(new    StatusObject("Printing    the
protein report for gene '" + gene.getFullName() + "'..."));
```

```
            try {
                final           Collection          tmp           =
appService.search(Protein.class, gene);
                final int size = tmp.size();
                if (size == 0) {
                    sb.append("No  proteins  found  for  "  +
gene.getFullName() + ".");
                    notifyObservers(STATUS_REPORT_DONE);
                    return;
                }
                sb.append("Protein name: ");
                for  (Iterator  iterator  =  tmp.iterator();
iterator.hasNext();) {
                    Protein      protein      =      (Protein)
iterator.next();
                    sb.append(protein.getName());
                    if (iterator.hasNext()) {
                        sb.append(", ");
                    }
                }
            } finally {
                notifyObservers(new      StatusObject("Protein
report done for gene '" + gene.getFullName() + "'!"));
            }
        }

        public void printGenes(Agent agent, StringBuffer sb) {
            notifyObservers(new StatusObject("Printing the gene
report for agent '" + agent.getName() + "'..."));
            try {
                final           Collection          tmp           =
appService.search(Target.class, agent);
                final int size = tmp.size();
                if (size == 0) {
                    sb.append("No  targets  found  for  "  +
agent.getName() + ".");
                    notifyObservers(STATUS_REPORT_DONE);
                    return;
                }
                sb.append(size + " targets found: ");
                Target[] targets = new Target[size];
                tmp.toArray(targets);
                for (int i = 0; i < targets.length; i++) {
                    Target target = targets[i];
                    printGenes(target, sb);
                    if (i + 1 < targets.length) {
                        sb.append("\n");
                    }
                }
            } finally {
                notifyObservers(new  StatusObject("Gene  report
done for agent '" + agent.getName() + "'!"));
            }
```

```
        }

        public void notifyObservers(Object o) {
            setChanged();
            super.notifyObservers(o);
        }

        public void printFullGeneReport(Gene geneFound, final
StringBuffer sb, int paragraphNumb) {
            sb.append(paragraphNumb + ". ");
            printGene(geneFound, sb);
            sb.append("\n-");
            printGeneOntology(geneFound, sb);
            sb.append("\n-");
            printProteins(geneFound, sb);
            sb.append("\n-");
            printGeneAliases(geneFound, sb);
        }

        public void printGenePathwayReport(Gene geneFound,
final StringBuffer sb, int paragraphNumb) {
            sb.append(paragraphNumb + ". ");
            printGene(geneFound, sb);
            sb.append("\n-");
            printPathways(geneFound, sb);
        }

        public void printGeneAgentCliTriReport(Agent agent,
final StringBuffer sb, int paragraphNumb) {
            sb.append(paragraphNumb + ". ");
            printGenes(agent, sb);
            sb.append("\n-");
            printAgent(agent, sb);
            sb.append("\n-");
            printClinicalTrials(agent, sb);
        }
    }
```

Coding JcaBIO: The CaBIOSearchEngine Class

CaBIOSearchEngine extends *Observable* to notify *observers* about what the search engine is doing so we can keep the users of the application informed about the status of the search. As described earlier, we provide two search capabilities in the SwingCaBIO application: one to create a Gene report and one to create an Agent report. We will call the Gene search method searchGenesWithGenePattern() and the agent search method searchAgentsWithAgentPattern() respectively.

The constructor for the class takes a `gov.nih.nci.system.applicationservice.ApplicationService` object that helps to run the initial search. Based on the caCORE API, we create an object called `GeneCriteria` and set the gene name with the pattern we're looking for. We run the `ApplicationService` and supply it with the object we need to retrieve. We then collect all the genes that match the input criteria and return the result in an array of `Gene` objects.

As with the `searchGenesWithGenePattern()` method, `searchAgentsWithAgentPattern()` returns an array of `Agent` objects found. The complete code for the `CaBIOSearchEngine class` is provided in **Listing 6.3**.

Listing 6.3. Class CaBIOSearchEngine

```
package org.jfb.jcabio;

import gov.nih.nci.cabio.domain.Agent;
import gov.nih.nci.cabio.domain.Gene;
import gov.nih.nci.cabio.domain.GeneAlias;
import gov.nih.nci.cabio.domain.impl.AgentImpl;
import gov.nih.nci.cabio.domain.impl.GeneAliasImpl;
import
gov.nih.nci.system.applicationservice.ApplicationService;

import java.util.ArrayList;
import java.util.Collection;
import java.util.List;
import java.util.Observable;
public class CaBIOSearchEngine extends Observable {

    private static final StatusObject STATUS_SEARCH_DONE =
new StatusObject("Search done!");

    private ApplicationService appService;

    public CaBIOSearchEngine(ApplicationService appService)
{
        this.appService = appService;
    }

    public    Gene[]    searchGenesWithGenePattern(String
geneNamePattern) throws SearchException {
        try {
            notifyObservers(new    StatusObject("Starting
search with gene pattern '" + geneNamePattern + "'..."));
    List resultList = appService.search(Gene.class, gene);
            Gene[] genes = new Gene [resultList.size()];
            resultList.toArray(genes);
```

```
                notifyObservers(STATUS_SEARCH_DONE);
                return genes;
            } catch (Throwable e) {
                notifyObservers(new    StatusObject("An    error
occured while searching for genes using gene pattern '"
                        + geneNamePattern + "'!", 10));
                throw new SearchException("", e);
            }
        }

        public    Agent[]    searchAgentsWithAgentPattern(String
agentPattern)
                throws SearchException {
            try {
                notifyObservers(new      StatusObject("Starting
search with gene pattern '" + agentPattern + "'..."));
        Agent agentCriteria = new AgentImpl();
        agentCriteria.setName(agentPattern);
        List   resultList   =   appService.search(Agent.class,
agentCriteria);
               Agent[] agents = new Agent[resultList.size()];
                resultList.toArray(agents);
                notifyObservers(STATUS_SEARCH_DONE);
                return agents;
            } catch (Throwable e) {
               notifyObservers(new    StatusObject("An    error
occured while searching for agents with gene pattern '"
                        + agentPattern + "'..."));
                throw new SearchException("", e);
            }
        }

        public void notifyObservers(Object o) {
            setChanged();
            super.notifyObservers(o);
        }
    }
```

`SearchException` and `StatusObject` respectively provide mechanisms to handle errors that occur during the search process and provide the user with messages on the status of the search. The code for these two classes is provided in **Listing 6.4** and **Listing 6.5** below.

Listing 6.4. Class SearchException

```
package org.jfb.jcabio;
public class SearchException extends Exception {
    private Throwable embedded;

    public SearchException(String s, Throwable throwable) {
```

```
            super(s, throwable);
            this.embedded = throwable;
        }

        public Throwable getEmbedded() {
            return embedded;
        }
    }
```

Listing 6.5. Class StatusObject

```
    package org.jfb.jcabio;

    public class StatusObject {
        private static final int NO_TIMER = 0;
        private static final String STATUS_TEXT = "Ready!";

        public static StatusObject STATUS_READY = new
StatusObject(STATUS_TEXT, NO_TIMER);

        private String statusText;
        private int timer;

        public StatusObject(String statusText, int timer) {
            this.statusText = statusText;
            this.timer = timer;
        }

        public StatusObject(String statusText) {
            this.statusText = statusText;
            timer=NO_TIMER;
        }

        public String getStatusText() {
            return statusText;
        }

        public int getTimer() {
            return timer;
        }

        public boolean hasTimer() {
            return timer != NO_TIMER;
        }
    }
```

Running the JcaBIO Application

As described in **Table 6.4**, among the caBIO domain objects, the gene object serves as central hub of the basic research objects and provides access to object such as organs, diseases, chromosomes, pathways, sequence data, etc. To begin with, therefore, we will create a Gene report using the `JcaBIO` application. **Fig. 6.4** shows the results of a gene report conducted to search for genes named "*erb*". Note that *wild-cards* (*) can be used for retrieving information on genes. In this case, for example, we have performed a search with *erb** which as the report indicates has identified genes called "*v-erb-b2 erythroblastic leukemia viral oncogene homolog 3 (avian)*", with the approved Human Gene Nomenclature Committe (HGNC) gene symbol *ERBB3* and "*v-erb-b2 erythroblastic leukemia viral oncogene homolog 2, neuro/glioblastoma derived oncogene homolog (avian)*" with the approved HGNC gene symbol *ERBB2*, both of which are members of a family of *growth factor receptor* genes called *epidermal growth factor receptors* (EGFR).

Fig. 6.5 displays the results of a pathway search for the keyword *erb**. The search identifies three genes that match the input keyword *erb**: *ERBB2*, *ERBB3* and *ERBB4*, the corresponding pathways the three genes are involved in and a link to the graphical representation of the pathways on the BioCarta website for each (as shown in **Fig. 6.6** for *ERBB2*).

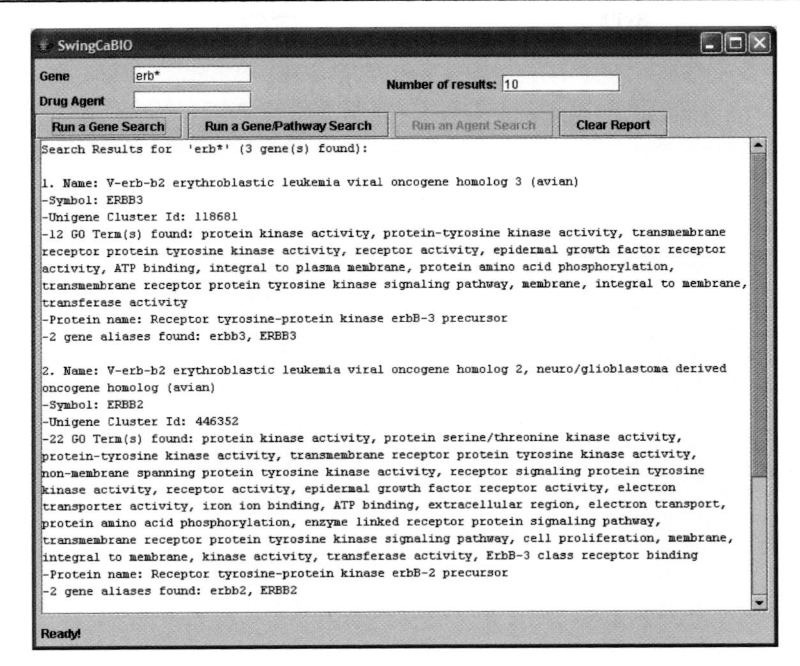

Fig. 6.4. Gene report for erb*

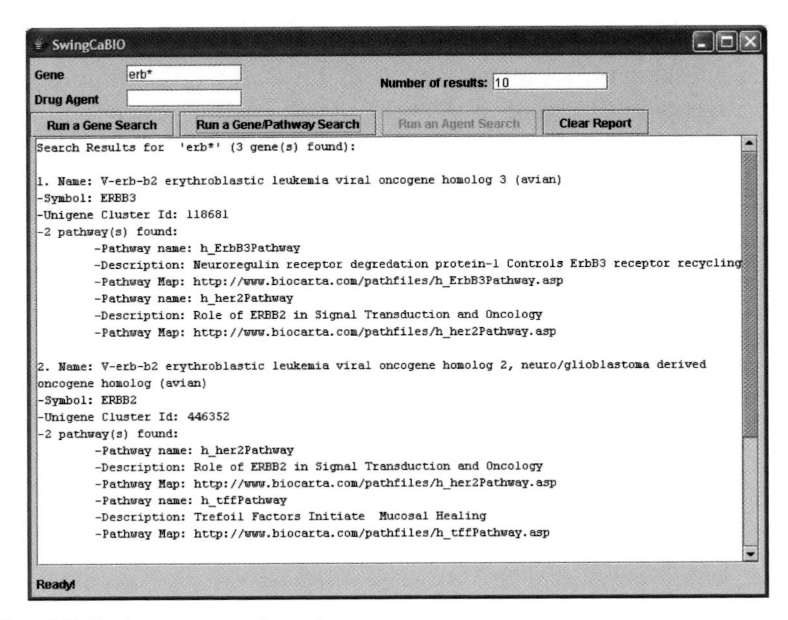

Fig. 6.5. Pathway report for erb*

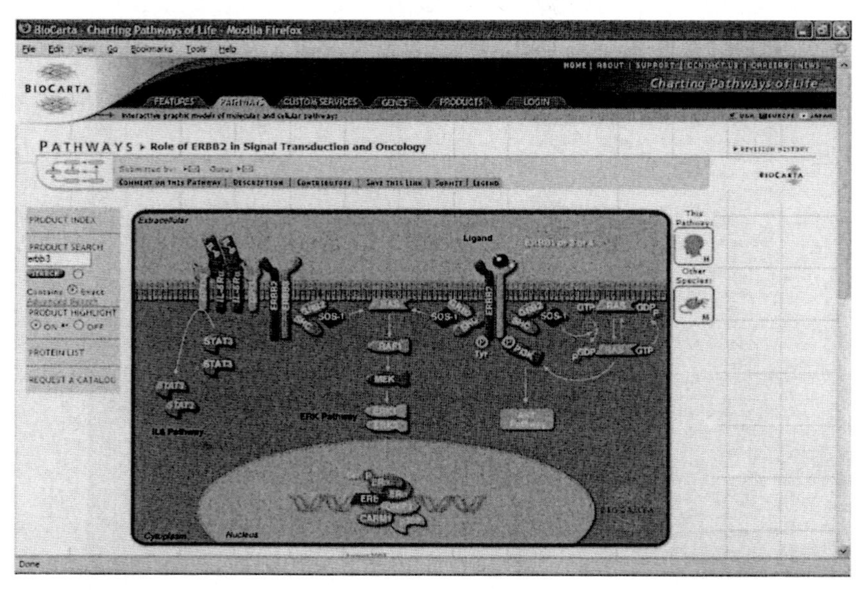

Fig. 6.6. Graphical representation of an *ERBB2* pathway in BioCarta

Next we will perform a therapeutic agent search for a well-known anti-cancer agent called *Taxol*. **Fig. 6.7** displays the results of a wild-card search performed with the term *TAX**. As expected, the search resulted in reports on Taxol, a compound present in the bark of the Pacific yew tree (*Taxus brevifolia*), which was later found to possess anti-cancer properties and approved for the treatment of ovarian, breast and non-small cell lung cancer. The report also presents detailed information on the many clinical trials that are being conducted using Taxol providing such details as the name of the study, its status, the organization conducting the study, Phase of the clinical trials and so on.

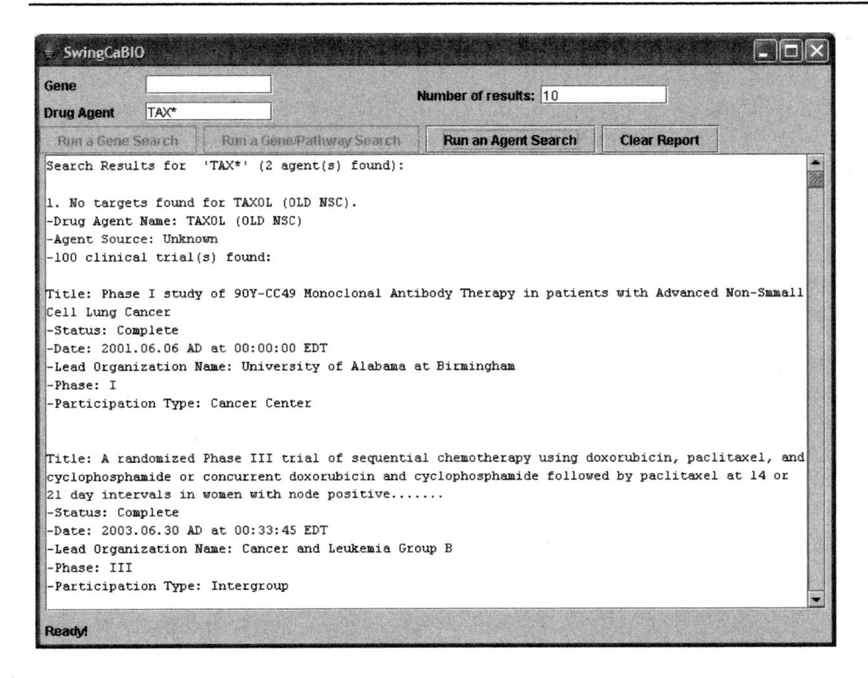

Fig. 6.7. Therapeutic agent report for Taxol

Summary

The NCI caBIG™ initiative is ushering a new era in cancer research by providing scientists with standardized tools to access and share information with one another overcoming cultural, geographical and technological barriers in ways not conceivable just a few years earlier.

In this chapter, we learnt about the rationale behind the creation of caBIG™ and the technologies that are being created or developed under the initiative to enhance the pace of cancer research. We created a very basic application to demonstrate a few of the many ways in which NCI's caCORE and caBIO domain objects can be used to retrieve information on biomedical objects in a way that bridges basic and clinical research. Needless to say, caCORE offers many more capabilities than what we have attempted to demonstrate and we encourage readers to take these small examples as a springboard to gain a better understanding of the power of the technology and build more complex queries as dictated by their individual research needs.

The power of the caBIG™ concept is uniting cancer researchers across the world. A similar initiative was launched by the UK National Health Service (NHS) for the development of cancer research informatics in that country through a strategic partnership with the NCICB on the caBIG™ effort. Both the initiatives will work together to build a truly global infrastructure for cancer research. These are indeed very exciting times for biomedical and clinical research and it is hoped that the joint efforts of people across the world will eventually lead to the demise of the scourge that we are battling.

As a living testimony of the work being done in this area, the NCI was recently awarded the 2006 Computerworld Honors 21st Century Achievement Award for Science for their accomplishment under caBIG™ Program. The Computerworld Honors Program was established to honor people or institutions who apply Information Technology for the benefit of society. Further information on the award is available at http://www.cwhonors.org/archives/2006/index.htm and https://cabig.nci.nih.gov/News_Folder/NCI_award.

Questions and Exercises

1. The NCICB has launched the Open Development Initiative (ODI, http://ncicb.nci.nih.gov/NCICB/infrastructure/open_dev_initiative) as an opportunity for biomedical researchers and bioinformaticians to contribute to on-going development efforts in the cancer domain. Explore the caBIO, caCORE and other ODI's of interest to you and think of ways you can participate in this effort.

2. The observation that, "Gene and/or protein X is significantly overexpressed in a specific cell population, tissue and/or in a laboratory model of disease Y" is that fundamental first indication of evidence that feeds hypothesis driven research into the biology and treatment of disease.

    a. What caBIO objects would you need to establish a causative link between biomolecules expressed in specific tissues (for example, cerebral cortex) and disease (for example, Alzheimer's disease)?

b. How would you extend the query to identify pathways that the biomolecules participate in and discover known chemical agents that selectively inhibit or modify events along the pathways?
c. Which caCORE data stores would you mine for such information?
d. Given that the ultimate aim of caBIG™ is to make biomedical and clinical data accessible via the grid, how would you design an application to take the information obtained above to locate appropriate tissue samples, patient cohorts and on-going clinical trials for further analysis and validation studies? What technical and non-technical issues would you need to address to build such an application?
e. Create an application expanding available caBIG™ technologies and data stores that will allow users to run such queries.

Additional Resources

Select NIH/NCI resources

- caBIO - http://ncicb.nci.nih.gov/core/caBIO
- caCORE - http://ncicb.nci.nih.gov/NCICB/infrastructure
- CaDSR - http://ncicb.nci.nih.gov/core/caDSR
- CaMOD - http://cancermodels.nci.nih.gov
- CMAP - http://cmap.nci.nih.gov
- CTEP - http://ctep.cancer.gov/
- CGAP - http://cgap.nci.nih.gov/
- CGAP GAI - http://gai.nci.nih.gov/
- EVS - http://ncicb.nci.nih.gov/core/EVS
- GEDP - http://gedp.nci.nih.gov

- MMHCC - http://mouse.ncifcrf.gov/
- NCI metathesaurus - http://ncimeta.nci.nih.gov/
- NCI thesaurus - http://nciterms.nci.nih.gov/NCIBrowser/Dictionary.do
- UniSTS - http://www.ncbi.nlm.nih.gov/entrez/query.fcgi?db=unists

Other biomedical repositories and resources

- BioCarta pathways - http://www.biocarta.com/
- Gene Ontology Project - http://www.geneontology.org/
- IMAGE Consortium - http://image.llnl.gov/

Standards and protocols

- ISO/IEC - http://www.standardsinfo.net/isoiec/index.html
- ISO/IEC 11179 standard - http://metadata-standards.org/11179/
- SOAP - http://www.w3.org/TR/soap/

ETL tools

- Kettle - http://www.kettle.be/
- Octopus - http://www.enhydra.org/tech/octopus/index.html

Selected Reading

The caCORE Software Development Kit: streamlining construction of interoperable biomedical information services. Phillips J, Chilukuri R, Fragoso G, Warzel D, Covitz PA. BMC Med Inform Decis Mak. 2006 Jan 6;6:2.

Covitz PA, Hartel F, Schaefer C, De Coronado S, Fragoso G, Sahni H, Gustafson S, Buetow KH. caCORE: a common infrastructure for cancer informatics. Bioinformatics. 2003;19:2404–2412.

Database resources of the National Center for Biotechnology Information. Wheeler DL, Barrett T, Benson DA, Bryant SH, Canese K, Chetvernin V, Church DM, DiCuccio M, Edgar R, Federhen S, Geer LY, Helmberg W, Kapustin Y, Kenton DL, Khovayko O, Lipman DJ, Madden TL, Maglott DR, Ostell J, Pruitt KD, Schuler GD, Schriml LM, Sequeira E, Sherry ST, Sirotkin K, Souvorov A, Starchenko G, Suzek TO, Tatusov R, Tatusova TA, Wagner L, Yaschenko E. Nucleic Acids Res. 2006 Jan 1;34(Database issue):D173-80.

Appendix

Apache Ant and Tomcat

A web app needs to be deployed on a web application server in order to be accessible through a web browser. We will use Tomcat - a free, open-source implementation of Sun's Java *Servlet* and *JSP* technologies developed under the *Jakarta* project at the *Apache Software Foundation* (http://www.apache.org/) – for this purpose. Building refers to the process of creating a *WAR* file from the application. Deploying the web app means the installation of the *WAR* file on the web app server (Tomcat). This is accomplished with the help of a make tool called Ant that is portable across platforms, and developed by the *Apache Software Foundation.* Apache Ant and Tomcat are both available from the *Apache Ant and Tomcat Software Foundation* web site respectively. We begin with the installation and configuration of the development tools - the Apache Tomcat server, and, the Apache Ant build tool that we need to create our application.

Downloading the Apache Tomcat server

Navigate to http://jakarta.apache.org/tomcat/ and click on Binaries under Downloads along the left bar (**Fig. A.1**).

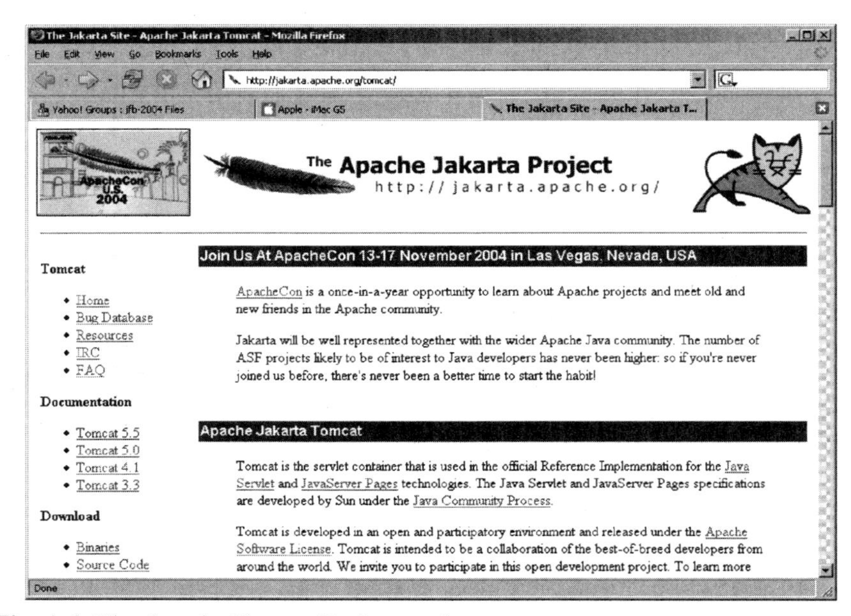

Fig. A.1. The Apache Tomcat Projects web page

This will take you to the Binary Downloads page (**Fig. A.2**).

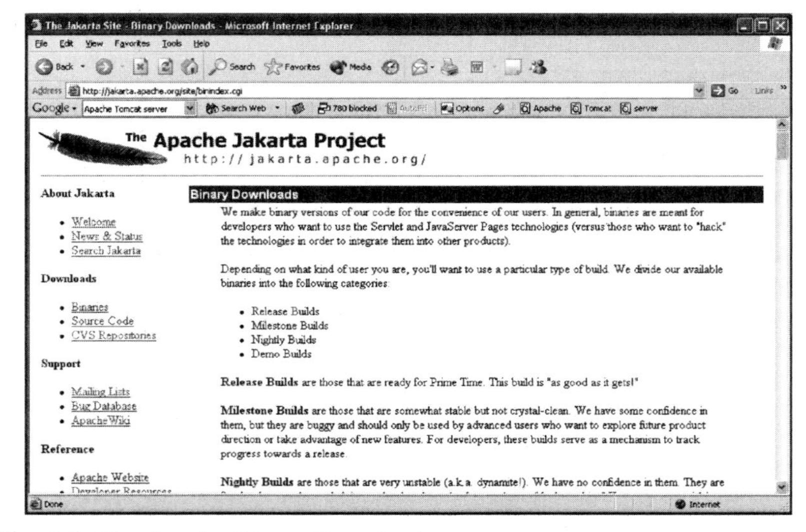

Fig. A.2. The Apache Binary Downloads page

Scroll down to where it says, “Tomcat 5.0.19” (at the time of this writing) and click on jakarta-tomcat-5.0.19.exe. This will download Tomcat-5.0.19 on your computer. Double-clicking on the downloaded file will open the Apache Tomcat Setup Wizard that will lead you through the installation process. For this installation, we are going to choose the C:\Program Files\Apache Software Foundation\Tomcat 5.0 directory. During the installation, you will be prompted to select a connector port and a password. Keep the port at 8080 and select a password of your choice and hit Next (**Fig. A.3**).

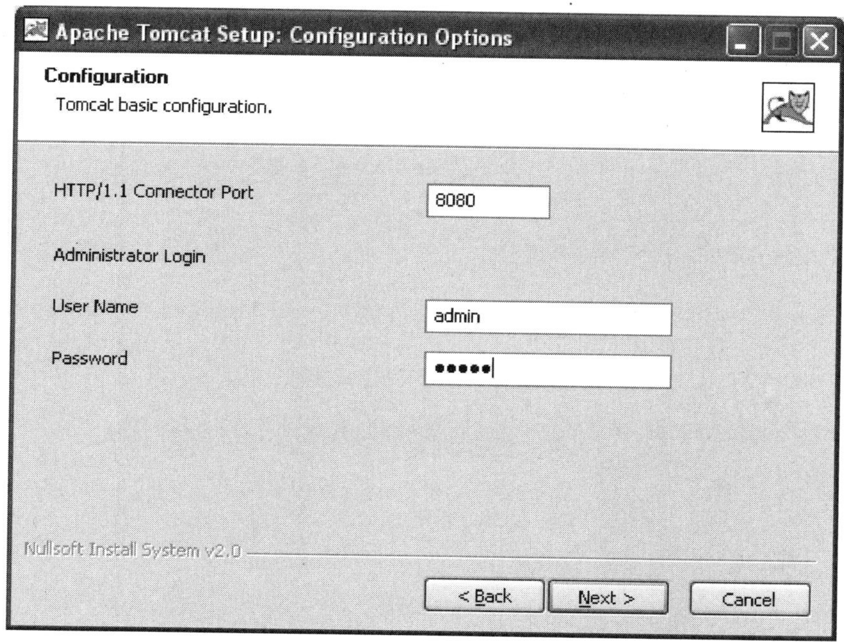

Fig. A.3. Apache Tomcat setup

You will also be prompted to enter the location of your *JDK* (*Java Development Kit*) installation because the Apache Tomcat Server is written 100% in Java and uses the compiler (provided in the development kit) to compile the *JSP* at runtime (**Fig. A.4**).

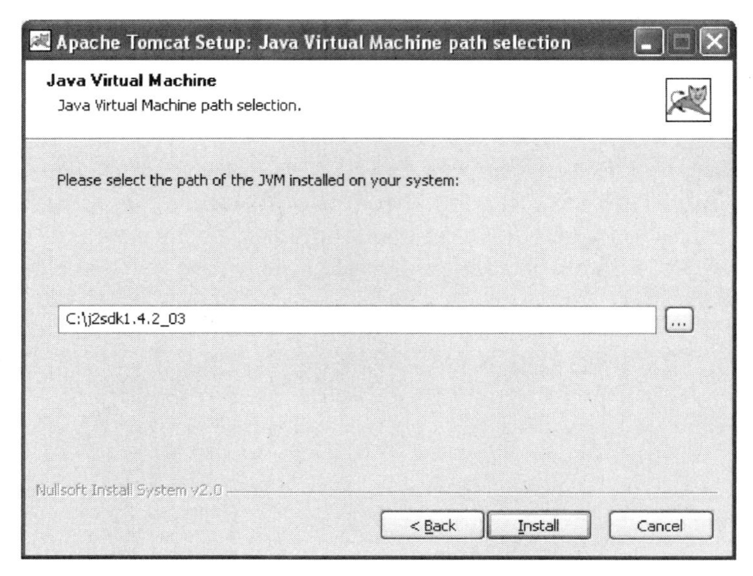

Fig. A.4. Apache Tomcat setup

Pressing Install will take you through the rest of the installation process. At the end of the installation, you will be able to start the server through the last screen by selecting the "`Run Apache Tomcat`" option (**Fig. A.5**).

Fig. A.5. Finishing Apache Tomcat Server installation

Pressing Finish will start the server and you will see the "Tomcat Webserver" splash screen for a few moments as the server starts up (**Fig. A.6**).

Fig. A.6. Tomcat Web App Server splash screen

You may also see a small icon on your Windows Task Bar at the bottom right indicating that the server is up and running (**Fig. A.7**).

Fig. A.7. Tomcat start and shutdown icon

You can open the Console Monitor by right-clicking on the icon and selecting the "`Open Console Monitor`" option. This opens a window listing out a log of the events that took place during start-up. You can shutdown the server by selecting the "`Shutdown:Tomcat5`" option. This removes the icon from the Windows Task Bar. To see the server in action, open up a browser window, type `http://localhost:8080` in the address bar and hit enter. This should open up the Tomcat home page and means that you have configured the server successfully (**Fig. A.8**).

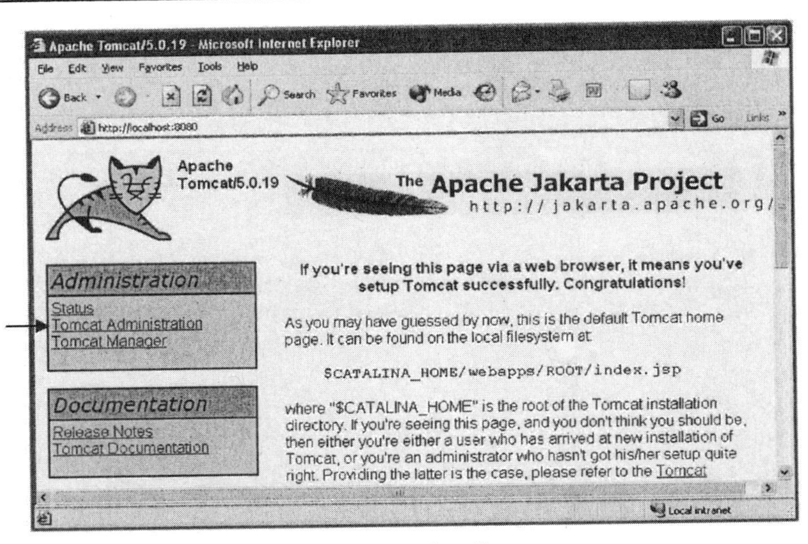

Fig. A.8. Testing Apache Tomcat Server on localhost

From this page, clicking on Tomcat Administration will take you to the Tomcat Webserver Administration page (**Fig. A.9**).

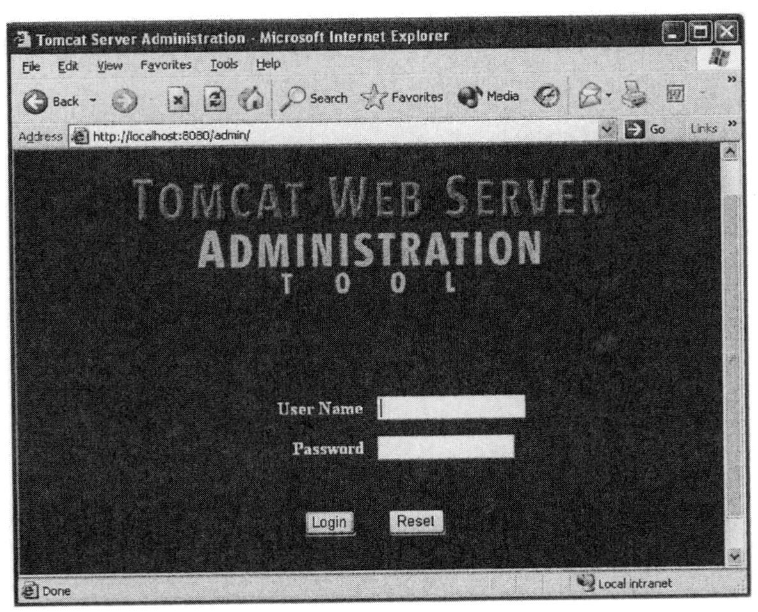

Fig. A.9. The Apache Tomcat Webserver Administration page

Enter the username and password you selected earlier to login (**Fig. A.10,** `admin` and `admin`, in our case).

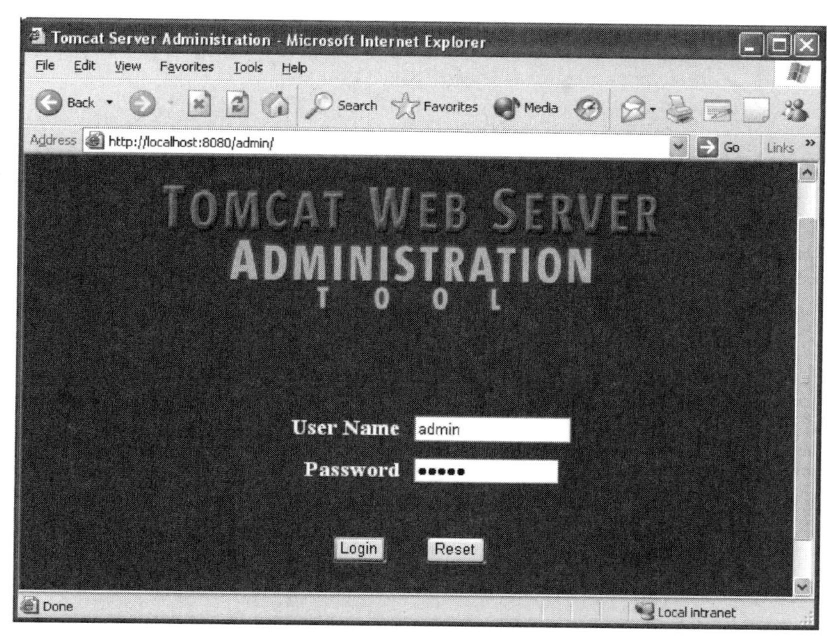

Fig. A.10. Tomcat Web Server log in

At this point you should see the Tomcat Web Server Administration Tool (**Fig. A.11**). The Tomcat Web Server Administration page essentially allows you to administer the server, that is, take care of issues such as assigning users, setting user roles (admin, manager etc) and privileges, also checking the services, resources and so on. For example, under Tomcat/Service/Host, the list of all the web app available on this server is displayed.

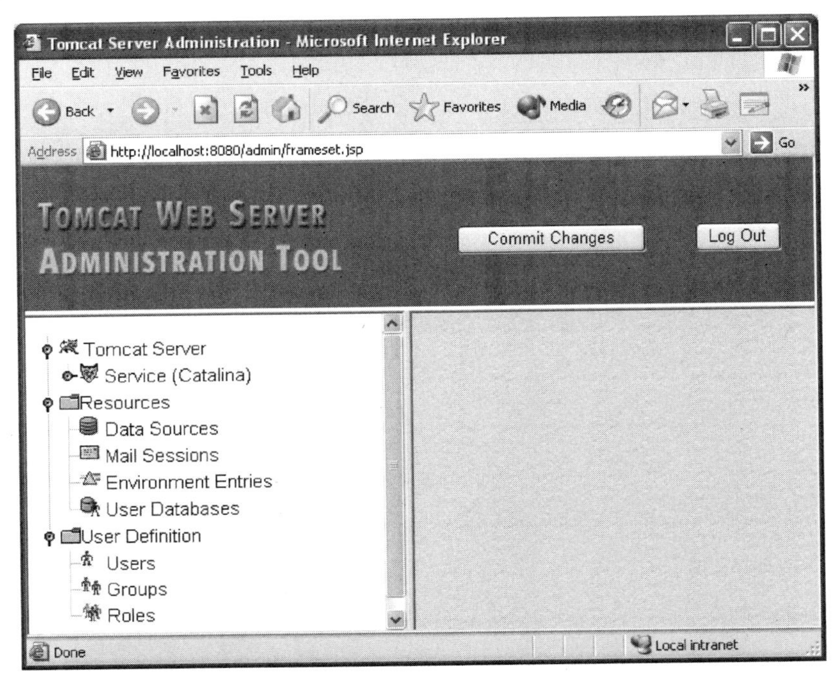

Fig. A.11. The Tomcat Webserver Administration Tool

Managing the Apache Tomcat Server

Let's create a new user called `manager`, enter a password and a Full Name (Tomcat Manager, this is an arbitrary string) and assign this user the role of a manager by checking the "`Manager`" check box under Role Name (**Fig. A.12**).

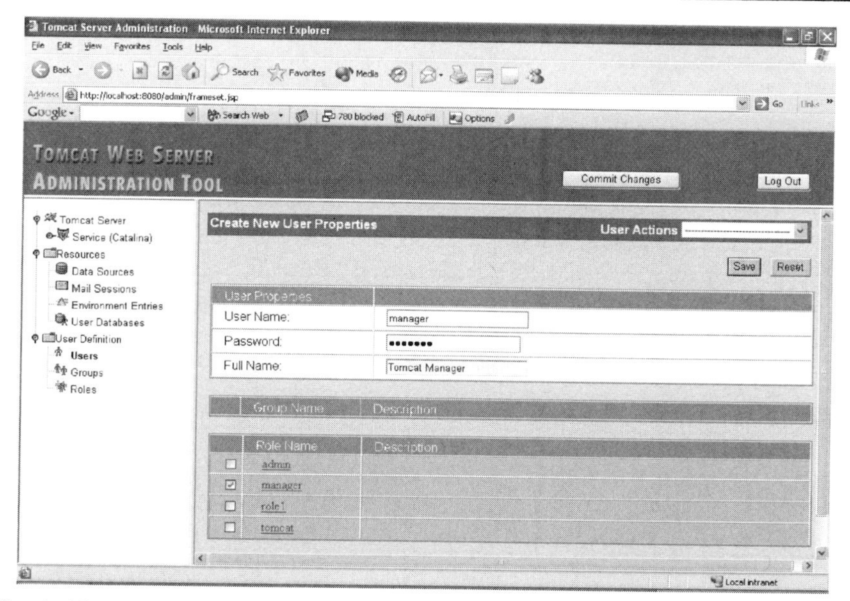

Fig. A.12. Assigning a new user

Pressing Save will add the user to the profile. Now lets try to login as manager by clicking Tomcat Manager. This will bring on the login window. Enter the username and password you just created and press OK (**Fig. A.13**).

Connect to localhost
Tomcat Manager Application
User name:
Password:
Remember my password
OK
Cancel

Fig. A.13. Tomcat Manager Application login

This will open up the Tomcat Web Application Manager (**Fig. A14**).

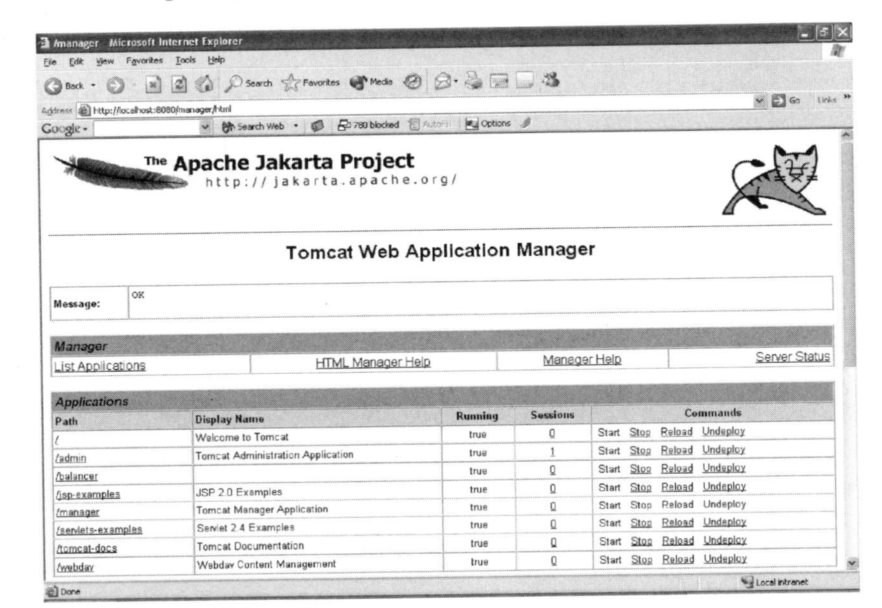

Fig. A14. Tomcat Web Application Manager

Starting the Apache Tomcat server

To start the Tomcat server, open the bin directory of your Tomcat installation (in our case, `C:\Program Files\Apache Software Foundation\Tomcat 5.0\bin`) and double-click on `startup.bat`. This will fire up the server (**Fig. A.15**). To shut down the server, double-click `shutdown.bat`.

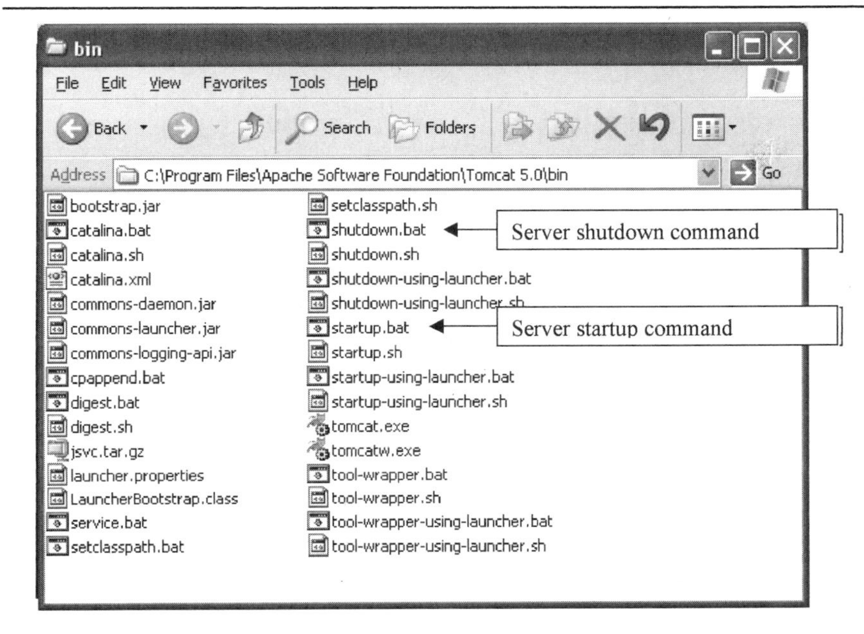

Fig. A.15. Starting Apache Tomcat Server

Starting the Tomcat Server brings on a DOS window that provides information on any error messages encountered during the start-up process. If everything went ok, it will say "`jk running` ..." and await for further instructions (**Fig. A.16**).

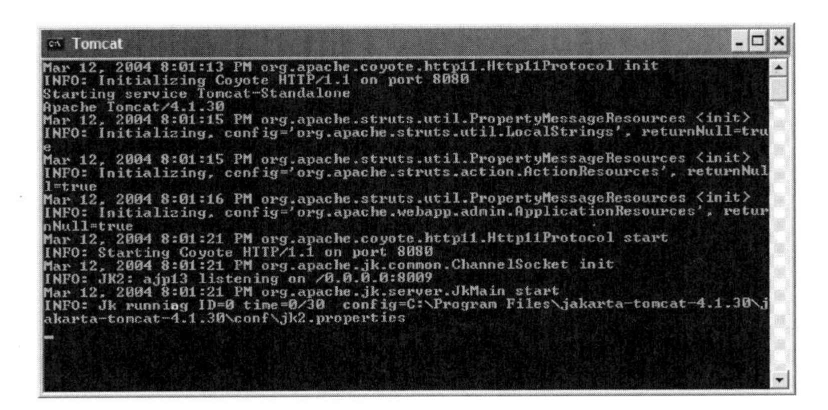

Fig. A.16. Starting the Tomcat Service

Installing and Configuring the Apache Ant Build Tool

Ant is a Java-based build tool like *Make*. It helps to automate tasks like creating directories, compiling, creating JARs, etc. Navigate to the Apache Ant Project website `http://ant.apache.org/` and click on Ant 1.6.1 (the latest version at the time of this writing, **Fig. A.17**).

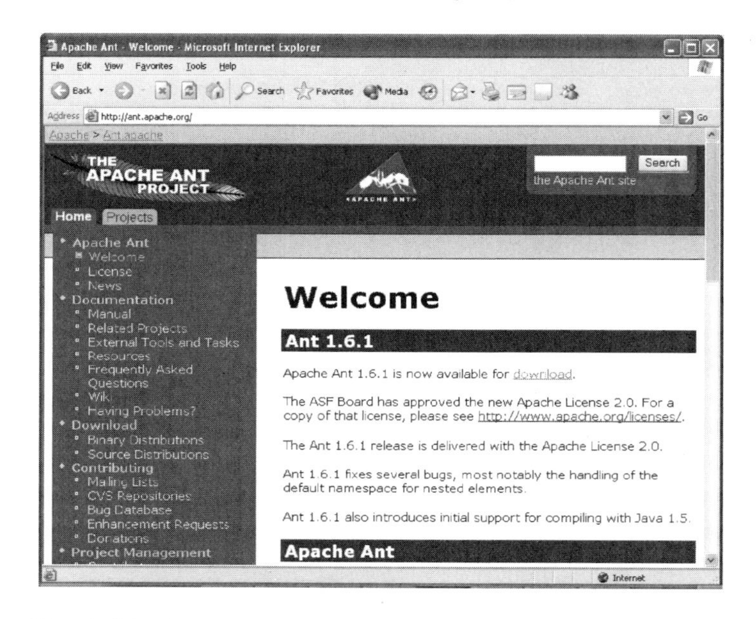

Fig. A.17. The Apache Ant Project website

This will take you to a page that lists the available Binary Distributions. Scroll down and locate the approach binary for your system (**Fig. A.18**), here we click on zip archive: `apache-ant-1.6.1-bin.zip`.

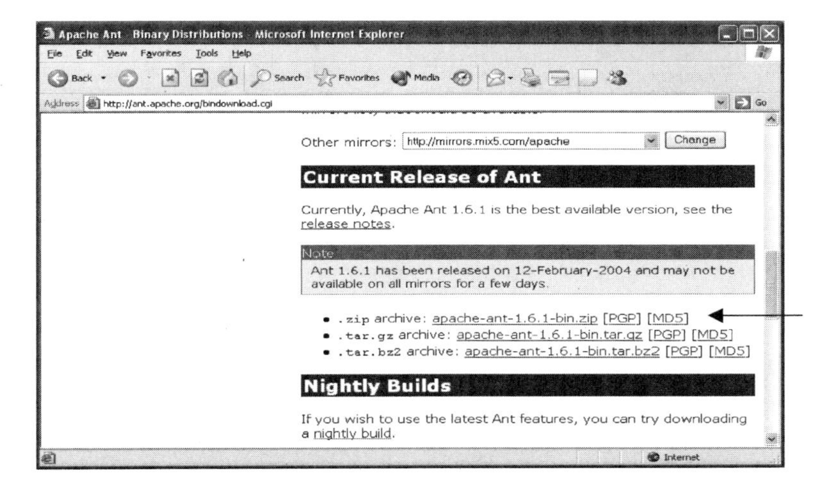

Fig. A.18. Ant binary distributions

Follow the instructions on the installation windows to install the Ant tool. Since this is a zipped file, you will need an unzipping utility to uncompress and extract the files. After this step, you should be able to see the installed files on your computer (**Fig. A.19**).

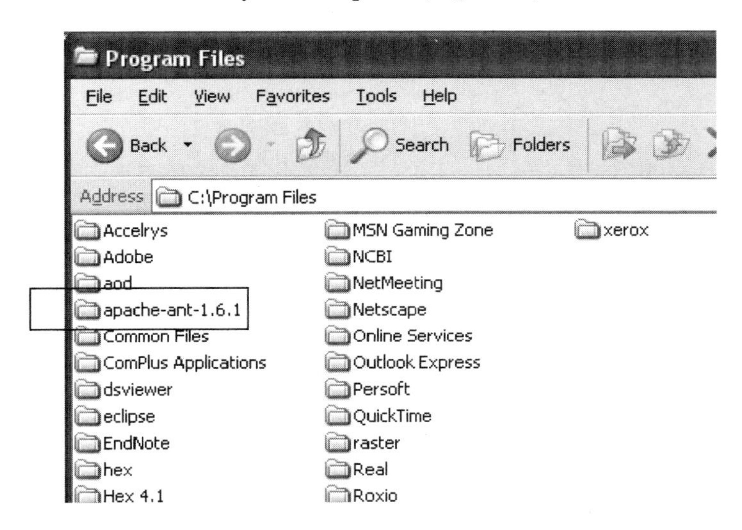

Fig. A.19. Locating installed Apache Ant files

The Ant utility is located in the `bin` directory (in our case, `C:\Program Files\apache-ant-1.6.1\bin`) (**Fig. A.20**).

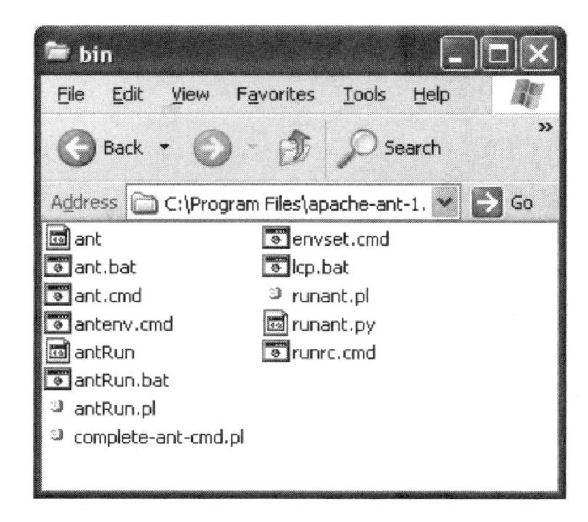

Fig. A.20. The Ant build tool

You should also be able to see instructions for using Ant by running the command:

```
ant –help
```

on the command-line from the Ant `bin` directory (**Fig. A.21**).

C:\WINDOWS\System32\cmd.exe

```
C:\Program Files\apache-ant-1.6.1\bin>ant -help
ant [options] [target [target2 [target3] ...]]
Options:
  -help, -h              print this message
  -projecthelp, -p       print project help information
  -version               print the version information and exit
  -diagnostics           print information that might be helpful to
                         diagnose or report problems.
  -quiet, -q             be extra quiet
  -verbose, -v           be extra verbose
  -debug, -d             print debugging information
  -emacs, -e             produce logging information without adornments
  -lib <path>            specifies a path to search for jars and classes
  -logfile <file>        use given file for log
    -l     <file>                   ''
  -logger <classname>    the class which is to perform logging
  -listener <classname>  add an instance of class as a project listener
  -noinput               do not allow interactive input
  -buildfile <file>      use given buildfile
    -file    <file>                 ''
    -f       <file>                 ''
  -D<property>=<value>   use value for given property
  -keep-going, -k        execute all targets that do not depend
                         on failed target(s)
  -propertyfile <name>   load all properties from file with -D
                         properties taking precedence
  -inputhandler <class>  the class which will handle input requests
  -find <file>           (s)earch for buildfile towards the root of
    -s  <file>           the filesystem and use it

C:\Program Files\apache-ant-1.6.1\bin>
```

Fig. A.21. Running the ant – help command

Make sure that you are in the right directory before you issue the above command. In this case, the correct path to `ant.bat` is `C:\Program Files\apache-ant-1.6.1\bin\` and so the command must be run from this directory. If this condition is not met, you will get the familiar DOS error message:

"`'ant' is not recognized as an internal or external command, operable program or batch file.`"

You will find information on Ant on the website that is simultaneously downloaded in your installation directory (for example, `C:\Program Files\apache-ant-1.6.1\welcome.html`).

Configuring environmental variables for Ant

To configure the environment variables for Ant on Windows, open:

Start→Settings→Control
Panel→System→Advanced→Environment Variables

and add the path information for Ant as shown in **Fig. A.22** and **Fig. A.23**.

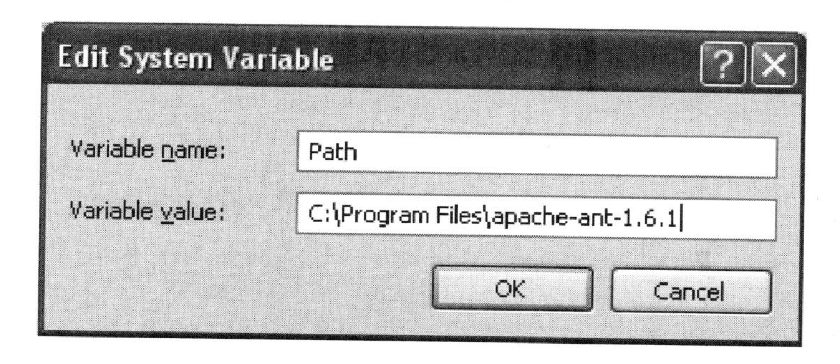

Fig. A.22. Setting the Ant Path

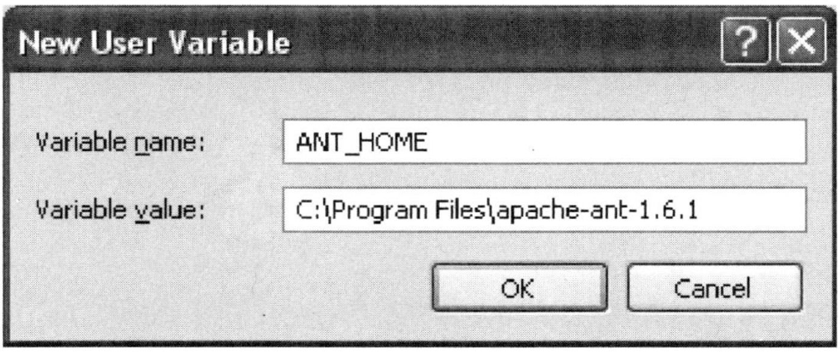

Fig. A.23. Setting the ANT_HOME constant

Building and Deploying The Web Application

Building the WAR file

Ant is a *Make* tool that is used to automate the creation of *WAR* files. It uses a *build file* written in XML usually called `build.xml`, although the file name can be changed. By default Ant uses the `build.xml` file that is located in the current directory where the user start Ant. The build file contains a series of instructions to Ant that define the processing required to successfully deploy the web application on the server. It defines what are known as "*targets*", which in turn run discrete tasks - pieces of code that can be executed independently – to compile the application and install it on a server. Additional tasks, for example, reloading a modified application onto a server or removing ("cleaning") older copies of the application to regenerate their content, can also be defined.

Let's assume the development directory has the following structure:

Src	The source directory for the Java code of the web app
Lib	The libraries used by the web app
resources	The resources like JSP, HTML, PNG, other images, that are used by the web app
web.xml	The web app deployment descriptor
build.xml	The build description file

The need for a build tool to deploy the application becomes obvious when one considers the large number of steps required to perform the same action manually:

1. Create "`dist`", the distribution directory
2. Create "`WEB-INF`" inside "`dist`", then the "`classes`" and "`lib`" directories inside "`WEB-INF`"
3. Compile the Java source classes inside "`dist/WEB-INF/classes`" directory
4. Copy the `web.xml` file into "`dist/WEB-INF`"
5. Copy the *JAR* libraries needed by the web app from the "`lib`" directory into "`dist/WEB-INF/lib`"
6. Create the *WAR* file using *Java jar tool*

This process is automated through the use of Ant. First, we create properties to let Ant know where to define resources; next, for each step we create an Ant target defining actions to perform. An example of the `PubMed` `build.xml` file is as follows:

```
  <project   name="PubMed   Project"   default="create-dist"
basedir=".">

  <property name="dist.home" value="${basedir}/dist"/>

  <target name="create-dist"
          description="Create binary distribution">
    <mkdir dir="${dist.home}"/>
    <mkdir dir="${dist.home}/WEB-INF"/>
    <mkdir dir="${dist.home}/WEB-INF /classes"/>
    <mkdir dir="${dist.home}/WEB-INF /lib"/>
  </target>

  </project>
```

Then by calling the Ant engine and the target we want to run, we will have the web app file structure created in the `dist` directory

```
ant create-dist or ant
```

Since we defined `create-dist` in the default attribute of the project tag, the second command line will run the same target as the first command.

Deploying the application on Tomcat using Ant

Tomcat provides a manager for web applications (installed by default on the context path `/manager`) that allows a user to deploy, install, reload, remove, start, stop any application or list all the applications available on the server from the web browser. Tomcat provides Ant tasks that can then be used inside a target in order to manage the web applications. This allows us to automate the deployment of `PubMed` web app by defining targets in the `build.xml` file as shown below:

```
  <project    name="PubMed    Project"    default="create-dist"
basedir=".">

  <property name="dist.home" value="${basedir}/dist"/>

  <!--These properties generally define file and directory
names (or paths) that affect where the build process stores
its outputs.

app.name                  Base name of this application, used to
                          construct
                          filenames and directories.

Defaults to "myapp".

app.path                  Context path to which this application
                          should be deployed (defaults to "/" plus
                          the value of the "app.name" property).

build.home                The directory into which the "prepare"
                          and "compile" targets will generate their
                          output.

Defaults to "build".

catalina.home             The directory in which you have installed
                          a binary distribution of Tomcat 5.  This
                          will be used by the "deploy" target.

dist.home                 The name of the base directory in which
                          distribution files are created.

Defaults to "dist".

manager.password          The login password of a user that is
                          assigned the "manager" role (so that he
                          or she can execute commands via the
                          "/manager" web application)
manager.url               The URL of the "/manager" web
                          application on the Tomcat installation to
                          which we will deploy web applications and
```

```
                    web services.

 manager.username   The login username of a user that is
                    assigned the "manager" role (so that he
                    or she can execute commands via the
                    "/manager" web application)

   -->

   <target name="create-dist"
             description="Create binary distribution">
     <mkdir dir="${dist.home}"/>
     <mkdir dir="${dist.home}/WEB-INF"/>
     <mkdir dir="${dist.home}/WEB-INF /classes"/>
     <mkdir dir="${dist.home}/WEB-INF /lib"/>
   </target>

   <property name="app.name" value="ncbi"/>
   <property name="app.path" value="/${app.name}"/>
   <property name="build.home" value="${basedir}/build"/>

   <property name="dist.home" value="${basedir}/dist"/>
   <property                                name="manager.url"
value="http://localhost:8080/manager"/>
   <property name="manager.username" value="tomcat"/>
   <property name="manager.password" value="tomcat"/>

   <!–

   These properties define custom tasks for the Ant build tool
that interact with the "/manager" web application installed
with Tomcat 5. Before they can be successfully utilized, you
must perform the following steps:

   - Copy the file "server/lib/catalina-ant.jar" from your
   Tomcat 5 installation into the "lib" directory of your Ant
   installation.

     -    Define    the    appropriate    values    for    the
"manager.password", "manager.url", and "manager.username"
properties described above.

     For more information about the Manager web application,
and    the    functionality    of    these    tasks,    see
<http://localhost:8080/tomcat-docs/manager-howto.html>.

   -->

   <taskdef                                      name="install"
classname="org.apache.catalina.ant.InstallTask"/>

   <target name="install" description="Install application to
servlet container">
```

```
    <install url="${manager.url}"
                  username="${manager.username}"
                  password="${manager.password}"
                  path="${app.path}"

              war="file:////${dist.home}/${app.name}.war"/>
  </target>

  </project>
```

The XML tag `taskdef` tells Ant to import the `InstallTask` Java class into the build space. The target named "`install`" can use it by defining the attributes needed by the task to support the install operation on Tomcat. In a platform independent way, we can deploy the `PubMed` web app on any web app servers available on any machine. This is in line with the "Write Once, Run Anywhere" Java principle.

Version Control Systems

There is no software development without a *Version Control System.* A project has a *life cycle*, where every little change that modifies the behavior of the application is important and needs to be documented. With version control systems, a developer can retrieve code written in the past for a feature that was removed because the project was missing some resources, but now needs to be put back because of the availability of new resources. Version control also makes it easier to maintain code when multiple developers are working together on a project. It is easier to integrate changes made by individual developers rather than exchanging the files by hand and incorporating changes using an editor.

Version control systems provide the functionality to compare one version of code with another, merge multiple versions of the same file, lock a file to avoid editing by other users while in use, access the modifications using web interfaces, etc. To use version control, the project team must first create a repository that it will use to manage the source files and code versions. Each member of the team then imports the source code files into the repository.

An example of a version control systems is *CVS* (*Concurrent Versions Control*), which is widely used open source environment. CVS allows users to import, commit, remove files, manage different code versions, create branches to develop patches, etc. Here is an example of how to import a project, check it out (retrieve a copy of the latest version of a file), work with it and check it back in the repository, on a Unix system.

First, we create a directory for the repository called `/cvs-repository` and we initialize CVS using the command `cvs -init`. Then we create a project directory called, for example, `swingblast`. To initialize and then import our project using CVS, we run the following command on the terminal:

```
cvs -d /cvs-repository init

cd swingblast

cvs -d /cvs-repository import -m "Important SwingBlast"

swingblast JFB INIT
```

The `import` command above tells CVS to use `cvs-repository` as the repository and to import the content of the current directory we are in, into the CVS directory `swingblast`, using the vendor tag JFB and INIT as the release tag.

Now, we back up our source code by running the following command:

```
tar —cvf swingblast-sav.tar swingblast
```

To check out the last revision of our project, we issue the following command:

```
cvs co swingblast
```

With these commands, we have achieved our first integration of the project in CVS. Another example of version control is Subversion. Subversion aims to create a more sophisticated tool than CVS. It uses most of the conventions used by CVS and adds new features like directories, copies, renames, truly atomic commits, network server options and efficient handling of binary files among other features.

Additional Resources

- Ant manual - http://ant.apache.org/manual/index.html

- Apache Ant download - http://ant.apache.org/bindownload.cgi
- Apache Software Foundation - http://www.apache.org
- Apache Tomcat - http://jakarta.apache.org/tomcat/
- CVS (Concurrent Versions Control) - http://www.nongnu.org/cvs/
- Subversion - http://subversion.tigris.org

Index